HOW THE ROSE GOT ITS THORNS

& Other Botanical Stories

HOW THE ROSE GOT ITS THORNS

& Other Botanical Stories

Dr Andrew Ormerod

greenfinch

Contents

Introduction

Do plants have feelings? If so, how do they communicate them, and how do other plants respond? What do plants do when they are too hot or too cold? How do plants defend themselves against attackers?

This book shows that the answers to these questions are many and various: the lives of plants are every bit as challenging as the lives of animals and humans, and even more diverse.

Take their drinking habits. All plants need water, and they've adopted a range of strategies to ensure they have enough of it. Some store it in their leaves, stems or roots, then hang onto it by insulating their leaves with waxy layers. Others open their pores only at night to prevent water loss in the heat of the day. Plants have adapted their foliage to suit a range of climates, generally producing very small leaves in harsh, dry environments and very large leaves in humid, tropical settings. Alternatively, some species hide from unfavourable weather by shedding their leaves or by withdrawing below ground level, where they keep their heads down until conditions improve.

The other thing all plants need is light: some grow tall to get as near as they can to the source of it, but those that lack supporting trunks use a range of structures to grasp, scramble or twist their way to the top. Some of these structures, such as prickles, spines and thorns, are dual purpose: they also defend the plants against predators.

Another question: is it better to live life in the fast lane and die young or to stick it out for the long haul over hundreds of years? In the plant world, the answer varies widely from species to species. Birch and poplar trees, for example, flower quickly, produce lots of seeds to perpetuate themselves, then wither away. This is the strategy of annual plants and pioneer trees – the earliest arrivals on previously open land.

By contrast, long-lived trees, such as oak, grow slowly and produce chemicals that prevent damage. They can live to a ripe old age, sometimes growing in harsh conditions, particularly when they are in the open, where they have the chance to stretch out and fully express their growth characteristics, rather than in forests. In their later years they pull their resources into their core, losing their top branches and reshooting from their bases.

Trees are responsive to external stimuli – when buffeted by the wind, for example, their root systems become tougher and stronger and their trunks become thicker – a bit like humans when they do regular exercise. A thickset tree is less likely to fall over in a gale than a tall thin tree that hasn't been honed by external forces. Grasses and legumes that grow in harsh conditions or are regularly grazed have low growing points so that no matter how hard the wind blows and no matter how many of their flowers are eaten by sheep and cattle, the plants themselves will survive because their vital parts are hidden away below where gales and teeth can reach them.

Many of the plants featured in this book have an adaptability that ranges from the surprising to the awe-inspiring. Alpine species originally evolved to survive in poor soil and under year-round battering by extreme cold. But they take to other conditions, too: they can thrive in temperate climates and thus have unexpected potential to cope with global warming. Other plants can prosper in settings that might be expected to kill them even faster than climate change: ferns can live on land contaminated by arsenic; sunflowers can cope with strontium and caesium, which are normally toxic. Humans are increasingly using these plants to detoxify soil that has been polluted by heavy industry.

Another fascinating group of plants are those that live on and off others. Some, such as mistletoe and yellow rattle, are parasites; others, such as orchids, do no damage to their hosts.

Some plants are highly successful colonizers: buddleia and dandelions produce light seeds that are dispersed on the air or in water and then establish themselves without difficulty in new locations. Other colonizers, such as bluebells and nettles, take advantage of woodland shade and fertile ground to become dominant species. A few species are mysteriously tied to a single location: the Italian bellflower grows naturally only on a small stretch of Mediterranean coast.

There is a quiet war going on in nature, and this book explores some of the ways plants protect themselves against assault by predators and herbivores. The deterrents include biochemical toxins and leaves with wax coatings and hairs. Many plants are geared to attract some insects and mammals – so that they can use them to spread their pollen – while repelling others, but some creatures draw strength from the very materials that are meant to poison them. The leaves of milkweed are toxic (see pages 78–9), but monarch butterflies not only eat them but also use the chemicals they contain to enhance their own orange coloration, which in turn deters creatures that might otherwise eat them.

There are so many surprising things happening in the plant world that it's probably futile to try to identify the most amazing. However, a strong contender must be the way acacia trees in Africa warn their neighbours of approaching herbivores so that their leaves have time to generate more of the tannin that the grazing animals don't like to eat (see pages 116–9).

Modern scientific techniques, such as molecular genetic assessment (genomics), have produced great leaps forward in our understanding of plants. This book highlights some of the most interesting findings. But there is still much to discover, much to learn about the mysterious and endlessly fascinating world of flora.

The
Stories

How the gorse got its shape

What is the true cost of your honey? The price may vary greatly, depending on whether it's a global brand (cheap) or a special local product (expensive). Plants pay a high energy price for pollen and nectar production, but that outlay is necessary to entice insects in to cross-pollinate their flowers.

The problem is that lots of creatures enjoy a plant's honey or nectar but do nothing for the plant in return. So some flowers have evolved to ensure that their pollen is accessible only to insects that will carry out cross-pollination. Gorse – part of the Faboideae subfamily of legumes, which includes peas and beans – is one of many plants whose flowers are papilionaceous (butterfly-shaped) and whose leaves are not symmetrical but dorsiventral – that is, with surfaces that differ from each other in structure and appearance.

Plants like these undergo physiological changes early in their development, morphing from symmetrical flowers such as buttercups into their own distinctive bilateral symmetries.

Gorse flowers are throat-shaped, but they don't swallow bees, they just tempt them in to take their pollen.

Gorse flowers are shaped in ways that encourage honey-bees to land on them.

Bees collect pollen and nectar separately and, in response to this, gorse and other specialist flowers produce very little nectar so as not to distract the insects from pollination.

Gorse flowers have five petals. One is a banner that stands up and attracts pollinating insects. Two are horizontal keel petals that are fused at their base and encase the male and female parts of the flower. These are flanked at the top by two wing petals that form a horizontal landing platform for pollinators.

When a bee or some other large pollinating insect lands on the platform, it is heavy enough to depress the wing and keel petals. This exposes the ripe anthers, which explosively release pollen onto the bee's abdomen while the stigma brushes the same part of the insect to receive any pollen that it may previously have acquired from other gorse flowers. The gorse flower can also be self-pollinated by bees or by small insects running around inside the flower. Among the other flower shapes that encourage bees to land on them are bilabiates (plants with two lips); examples include snapdragons and mints.

They say that when gorse is out of flower, kissing is out of fashion! That is because you can find gorse in flower all year round in its natural habitat. There is a main flowering season when the air can be suffused with a perfume reminiscent of coconut suntan cream. Some gorses flower through the winter months in order to increase their chances of survival: winter flowers are less likely to be damaged by mites or other insects laying their eggs on developing gorse pods and seeds, so this provides an insurance policy for the main season when the bulk of seeds are produced.

Some ornamental and crop plants, such as sweet peas, peas and field beans, have flowers that are too large for pollinating insects to depress the keel petals. Bees and other insects have developed methods of overcoming this difficulty: they have learned to cheat by biting holes in the back of the flowers to search for nectar and pollen without triggering the keel petal platform designed to facilitate cross-pollination.

Campanula flowers are mostly bell-shaped, but sometimes they open out to look like flat stars.

How the campanula got its hump

Campanula are some of the most popular easy-to-grow plants. There are more than 500 species, which thrive in a variety of locations and climates throughout the northern hemisphere.

Campanulas, which are generally bell-shaped, include a wide variety of small alpine species that have adapted themselves to live in cracks in rocks in isolated mountain habitats. Examples include the Olympic harebell (*Campanula piperi*), so named because it is found on the Olympic Peninsula in Washington State, USA, where it benefits from summer melt-water that runs off nearby glaciers.

Many alpine species living in rock cracks are deep-rooted, short and squat to minimize wind damage. They grow in clumps that may look like little humps in the soil. By contrast, the creeping bellflower (*Campanula rapunculoides*) has adapted itself to semi-shaded areas, such as open woodland in Europe and Siberia. Its roots penetrate deep into the ground to access nutrients and water, making it a hardy and adaptable plant. It is good-looking, but nevertheless you don't want it in your garden: on introduction to the USA, it proved to be an invasive weed that drove out competing flowers.

In striking contrast to the small alpine species, the giant bellflower (*Campanula latifolia*) can grow up to 1.8 m (6 ft) tall. It has large, bell-shaped flowers that come in various shades of blue, purple or white. It is native to Europe and western Asia, where it has adapted naturally to shady environments – broad-leaved woodland, coppices, parkland and the edges of forests – by growing quickly and vertically to get maximum benefit from the limited light available.

Found throughout Europe, North America and Asia, the harebell (*Campanula rotundifolia*) has a genetic complexity that enables it to thrive in a wide range of habitats and soil conditions. Rarely the dominant species anywhere, it prefers open, sunny land and upland soils. It is ideally adapted to grow in closely grazed, dry or drought-prone areas because of its taproot (a large, central root from which other roots sprout laterally). Harebells are less likely to flower in woodland than in open habitat, and are not normally found on fertile lowland soils.

Various forms of harebell have developed that can tolerate sites polluted by heavy metals, but such growths tend to be smaller than would be expected in uncontaminated locations. Different forms have been collected for trials in the UK and Europe so that scientists can examine the plant's variations in growth and its potential to adapt to climate change.

Although most campanulas are highly adaptable, there is
at least one outstanding exception to this general rule.
The native habitat of the Italian bellflower (*Campanula isophylla*)
is a small stretch of land near Capo Noli on the Mediterranean
coast of Italy, and the plant is found nowhere else, despite
its flowers readily producing small seeds capable of wind
dispersal. No one quite understands why it has failed to spread
and establish itself elsewhere. Yet despite its limited natural
distribution, this species has become enormously popular as a
potted plant for hanging baskets and house plants all over the
world, particularly in Scandinavia. One of the reasons for its
success in this market has been the discovery that its growth
pattern (which is the same as that of several other *Campanula*
species) readily adapts itself to variations in day and night
temperatures. If days are warm and nights are cool, the plants
produce longer stems: this growth spurt occurs at the end of the
night. In the 1980s, US horticulturists developed methods of
manipulating glasshouse temperatures to accelerate this growth
and thereby increase the supply of Italian bellflowers to florists
and garden centres: thus commerce has enabled this charming
little plant to spread much further from its Riviera roots than it
would ever have managed if it had been left to its own devices.

How the ranunculus got its skin

How many mouths does a plant have? Humans have one mouth for gaseous exchange and to take in sustenance. Plants have many mouths in the form of stomata – small openings on their leaf surfaces.

Stomata allow the plant's leaves to take in carbon dioxide, the feedstock for photosynthesis. When they absorb water, they become swollen and their cell walls bow out, causing the stomatal pores to open. Environmental stimuli, such as light, temperature, humidity, carbon dioxide levels in the air and the presence of plant pathogens, may also affect stomatal opening and closing.

Dicotyledonous plants (flowering plants that have pairs of leaves) tend to have more stomata on the undersides of their leaves because these are cooler surfaces. By contrast, monocots (plants with only one seed leaf) such as grasses and bulbs tend to have equal numbers of stomata on both leaf surfaces, while aquatic plants have stomata on their upper leaf surfaces only.

The leaves of the garden ranunculus have adapted to prevent water loss.

To prevent excess water loss, some plants, such as sedums, open their stomata only at night to store carbon dioxide for photosynthesis through a process known as crassulacean acid metabolism (CAM). This is found in plants adapted to dry conditions. Other leaf structures reducing water loss include a waterproof cuticle (a protective film covering the outer skin layer of leaves) and trichome hairs (very fine outgrowths) on leaves, which reduce water evaporation.

So how did stomata originate? Around 400 million years ago primitive plants such as algae lived in water and didn't need stomata to control loss of fluid. Stomata evolved when plants such as mosses came to live on land. Some land plant species have adapted to be amphibious to cope with alternate flooding and dry conditions. A leading example is the pond water crowfoot (*Ranunculus peltatus*), which has thin and feathery leaves beneath the surface to absorb nutrients, while its leaves

Sedums are succulents – plants that can store water and nutrients in their leaves.

above the surface have a skin-like quality and purpose: they are thick and robust to provide support and prevent water loss.

This process of producing different types of leaves on the same plant in response to changing conditions is termed heterophylly. In water cabbage (*Rorippa aquatica*), the plant's submerged leaves have gone a stage further and don't produce stomata, while its aerial leaves still retain them. Even more extreme, eelgrass (*Zostera marina*) has returned to the sea and lost all its specialist stomata control mechanisms because it no longer needs them.

Stomata are generally evenly spread over the leaf surface with always at least one cell between them to enable their guard cells (protective coverings) to open and close properly. In dry climates, some sedums and begonias have concentrated clusters of stomata rather than an evenly spread pattern to keep the water concentrated and thus minimize its loss.

Inevitably, some moisture is lost through the stomata of leaves, and that loss may play a role in enhanced localized rainfall patterns over inland forests. Rising atmospheric carbon dioxide caused by climate change reduces the number of stomata produced on the leaf. Understanding the control of stomata formation may influence the future efficiency of our food plants. For example, reducing stomatal density can increase water use efficiency in rice plants without significantly reducing their yield: this may help us to cope with future rising temperatures and drought conditions.

The leaf patterns of Dieffenbachia *may be attractive to humans but to other species they are a warning and a deterrent.*

How the leopard lily got its spots

The palette of bright colours on variegated leaf plants is loved by gardeners and flower arrangers alike. Patterned plants break the monotony of green leaves and provide delightful splashes of colour in gardens, houses and offices.

Flower colour patterns can include bicolour (spots) and venation (veins on leaves). The common names of coloured plants can be based on their physical traits: an outstanding example is the leopard lily, any of several unrelated species with spots and patterns on their leaves or flowers, such as forms of *Dieffenbachia maculata* and *Lilium pardalinum*. Although spots and patterns on garden and houseplants are attractive to humans, that is not the reason why these ornamental features occur in nature: ornamentation on leaves and flowers can provide protection from predators as well as attracting pollinators.

The leaves of *Dieffenbachia* are toxic: poison is the ultimate deterrent to predators. However, the patterns and spines on the surface of the leaves provide the first line of defence, the former as camouflage, the latter as a weapon. Patterned leaves can offer a survival advantage over green leaves, even though they reduce the plant's ability to capture energy from sunlight. For example, *Caladium* – a related foliage plant genus with a wide range of patterns and colours on its showy leaves – is not eaten in the wild by a type of moth called the mining moth. Research has shown that these moths prefer plants with green leaves, rather than plants with patterned leaves, because patterned leaves resemble recent damage caused by larvae. Green-leafed plants that are artificially variegated are avoided by these moths because the surface patterns signal to them that the leaves have already been eaten, so are not worth laying their eggs on.

The leopard lily (*Lilium pardalinum*) has orange-red to crimson Turk's cap (turban-shaped) flowers with maroon spots. If you look at them closely, you may notice that some have papillae (flowers with bumps), spots and spattered patterns of pigments, rather as if paint has been flicked onto them. Some spots may have haloes around them where there are either no, or reduced, pigments present. Such variations in patterns on the flowers are important for breeding new varieties of cut flowers and lilies for the garden, and geneticists are learning more about their origins. In the case of the beetle daisy (*Gorteria diffusa*) from South Africa, scientists now understand how the cells for flower pigments, raised petal surfaces and fine, hairlike appendages known as tricomes combine to form spots on the petals that mimic the appearance of the flies that pollinate them.

Spots and patterns have evolved independently on a wide range of flowering plant species, so there must be a natural reason for this, but it has often proved difficult to determine their intended role. It has been suggested that such patterns may attract pollinators – in addition to the known factors that attract insects to pollinators, such as pollen, nectar, scent and colour. Although *Lilium lancifolium* – a Turk's cap lily with orange flowers – lacks nectar or scent, it has been suggested that its pronounced flower spots and sticky pollen attract insects and help it to pollinate.

Scientists have tried to measure the relative attractiveness to pollinating insects of flowers with and without spots, but that has proved difficult because of other potentially influential factors that cannot be quantified, such as variations in nectar or aroma that may affect the results. Experiments with the common yellow monkey flower (*Mimulus guttatus*), using artificial flowers with and without spots, have tended to indicate that spots and patterns do attract pollinating insects.

The bergenia that had elephant's ears

Why do some plants have large leaves? Large leaves provide plants with a greater surface area for photosynthesis to capture sunlight for energy generation and transpiration for cooling. They also shade out competing plants.

Bergenias come from the high mountains of central Asia – some can tolerate temperatures as low as -35° C (-31° F). There are ten species and many garden-worthy hybrids. Two of the more common species in cultivation are *Bergenia cordifolia*, with heart-shaped leaves, and *Bergenia crassifolia*, with thick leaves. These tough, drought-resistant plants produce rosettes of evergreen leaves that have earned them nicknames including 'pig squeak', 'elephant ear', 'heartleaf', 'leather cabbage' and 'picnic plate'. They have beautiful pink to white flowers.

Bergenias contain phytochemicals that protect them from predation and cold weather. *Bergenia crassifolia*, also known as Mongolian tea, is used in traditional medicine to treat a range of illnesses, from headaches to haemorrhages.

When bergenias finish their upward growth, subsequent flowers sprout side-ways in clusters known as cymes.

Bergenias are highly adaptable but are often confined to the shade. However, it is in sunlight that they maximize their amazing range of autumn and winter leaf colours – bronzes, reds and purples produced by tannins. Their large leaves also shade out competition.

In general, throughout the world, the plants with the largest leaves are native to moist tropical climates. The plants with the smallest leaves are found in arid areas – mountains and towards the poles at high latitudes. Although sunlight and water are important factors in determining leaf size, the critical factor is temperature, especially at night.

For example, the leaves of banana trees are enormous: this is to enable the plant to prosper in hot, humid conditions. They are also relatively thin, to maximize photosynthesis. In bright light, small narrow leaves can intercept more light than more heavily shaded large leaves. However, larger leaves run a greater risk of heat and water stress in hot, dry environments.

The biggest and one of the rarest banana species is *Musa ingens*, the giant highland banana, from Papua New Guinea. It is the largest herbaceous plant in the world and six to seven times the size of the average banana plant. Its stem is made from the largest folded leaf stalk in the world at 15 m (49 ft) in height, and it produces leaves 6 m (20 ft) long – around a million times bigger than a common heather leaf. The total height of the plant, including the leaves, is at least 20 m (65 ft). Even its fruits are big: they grow in clusters weighing up to 60 kg (132 lb). There

*The banana plant
has the largest
undivided leaves
in the world.*

arc other tropical palms and bananas with leaves over 3 m
(10 ft) long – these dimensions far exceed those of the largest
leaves of temperate garden plants such as *Gunnera manicata*,
a giant rhubarb from Brazil, which reach a comparatively modest
2 m (6 ft) in diameter.

(It should be noted that these record sizes refer only to
undivided leaves: there are even larger growths, such as those
of the Amazonian palm [*Manicaria saccifera*], which may be up
to 8 m [26 ft] long, but they are divided at the tips.)

The giant miscanthus
(Miscanthus x giganteus)
makes magical sounds when
blown by the wind.

The miscanthus that sang to itself

If you listen to the sounds from the fields, woods and gardens when it's windy or raining, some of the noises you'll hear are generated by the weather, but others are produced by the plants and trees themselves.

When rustled by the wind, silvergrass (*Miscanthus*) – a widely distributed plant group in eastern Asia and North and South America – sounds like a bead curtain being pushed aside. Two species of poplar worth listening out for are the Frémont cottonwood (*Populus fremontii*), the sound of which has been likened to that of flowing water when rain hits the leaves, and the quaking aspen (*Populus tremuloides*), which is particularly appealing because of its rustling leaves. Both are native to southern central USA and northern central Mexico. In Europe, hedges of beech (*Fagus sylvatica*) make delightful sounds in the wind as they retain their dry leaves in the winter, while oak trees (*Quercus robur*) may hiss and sigh in the breeze. The calming effect of the sound of wind playing with the branches and leaves of plants has its own special name: psithurism.

Dried seed heads also make various noises when buffeted by the wind. Among the most striking sound effects are those produced in exposed locations by the stems of poppies (Papaveraceae), nigella (*Ranunculaceae*), foxgloves (*Digitalis*), greater quaking grass (*Briza maxima*) and yellow rattle (*Rhinanthus minor*). Some seed heads, such as those of the gorse family (*Fabaceae*), can be heard popping open to spread their seeds when ripe on warm sunny days.

This orchestra of sound is made possible because of the internal structure of these plants, which enables them to resist being bowled over by the wind. Their cells contain a variety of materials that not only give them strength but also contribute to the sound effects created. Among these components is silica – a hard mineral present in leaves – which provides stiffness and rigidity and contributes to the rustling noise as the leaves rub against each other.

Also present in plant cell walls is cellulose, which provides structural support and plays an important role in their mechanical strength – we encounter cellulose every day in paper and cardboard: think also of the distinctive noise created by scrunching cellulose bags! Another important contributor to the mechanical strength and rigidity of herbaceous plants and grasses is lignin, a hard and rigid polymer.

In addition to the noises generated by the action of external forces, some plants make sounds of their own. In dark rooms where rhubarb is grown commercially you may sometimes hear cracking and popping as the plants undergo rapid stem elongation. Rhubarb can even make noises when cooked: steam can build up rapidly in the stems, creating a popping sound.

In the 2010s, scientists in Israel and the USA placed tomatoes and tobacco plants in soundproofed acoustic boxes rigged up with ultrasonic microphones and recorded the sounds they made under various conditions. The plants that were well treated – fed and watered as they might expect – remained silent, but those that were distressed – either by being starved of water for up to five days or by having their stems cut – gave off noises. These sounds were between 20 and 250 kHz, so under normal circumstances inaudible to human ears, which can pick up only around 16 kHz, but once converted to audible wavelengths they were similar to the noises made by airlocks in central heating systems or the sounds of popcorn being cooked. The scientific name for noises of this kind is cavitation, and there is general agreement that in these cases they are expressions of distress.

The beginning of the amaryllis

Some plants have adapted to survive harsh environmental conditions by dying down to the ground and conserving energy and water in subterranean storage organs until favourable growing conditions return.

Storage organs include: taproots (as in dandelions); stem and root tubers (such as potatoes); corms (swollen stems on basal plates producing roots, as in crocuses); rhizomes (modified subterranean plant stems, such as ginger); and bulbs (fleshy leaves on short stems or basal plates: tulips, for example).

True bulbs are tunicate – that is, formed in concentric layers with a dry, papery covering outermost. Among them is the hyacinth. Imbricate bulbs consist of fleshy scales without a dry outer layer, for example lilies.

The fleshy leaf scales within bulbs on basal plates (stems) contain starch, proteins and other nutrients surrounding a growing point that develops into new leaves and flower stalks.

All amaryllis grow from bulbs – storage resources that enable the plants to survive poor soil and inclement weather.

During their dormant period, bulbs avoid being eaten by animals by dying back to the ground. Bulbs also contain defence compounds of varying potency to prevent them from being eaten. For example, although onion and garlic bulbs are edible by humans, their flavours are disliked by many animals and so prevent them from being devoured.

Bulbs can reproduce vegetatively by producing subdivisions at the base of the mother bulb. There are also several techniques for raising bulbs from fleshy leaves in dormant bulbs: these are used for propagating hyacinths and hippeastrum.

The belladonna lily (*Amaryllis belladonna*) originates from rocky areas in the Western Cape of South Africa. It is drought-tolerant and can grow into large flowering clumps from bulbs in open ground, subdivide and spread via wind-distributed seeds. It germinates rapidly in the spring in warm temperate climates. During dry summers, its foliage dies down to the ground, thus protecting the bulb from being grazed. Wildfires clear away dense vegetation, enabling overgrown colonies of bulbs to flower freely again. The bulbs produce beautiful, fragrant pink flowers on naked tall stems, which are pollinated by hawk moths at night and carpenter bees in the daytime. The long, strap-shaped leaves emerge after flowering has finished. The belladonna lily is a beautiful garden flower for the autumn, particularly in warm temperate gardens around the world.

The natural range of the sweetly scented garden hyacinth (*Hyacinth orientalis*) is around the borders of Iran, Iraq, Syria and Turkey, where it grows in the hills and mountains in scrub on limestone slopes, screes and cliffs. It has evolved from a wild blue-purple flower with around a dozen flowers per stem. Unlike amaryllis, its dormant period is during cold winters when it survives underground until the snows melt. The Romans seem to have spread the plant to the coast of France. Later, the Flemish botanist Carolus Clusius accelerated its spread throughout Europe. In the mid-1700s the discovery of double bulbs inspired a 'hyacinth mania' similar to the famous craze for tulips that had gripped Europe in the previous century.

That was a short-lived fad, but in the 21st century British grower Alan Shipp tracked down many of the old varieties, some from gardens in former Soviet countries. Perhaps the most remarkable rediscovery was that of 'Gloria Mundi' in a garden in Romania. First recorded in 1767, this hyacinth with white flowers and a red double centre has survived and regrown every year in obscurity because of its bulbous storage organ.

In the hands of green-fingered horticulturists, bulbs can survive for centuries as long as they are kept clear of pests, disease and viruses and fed correctly.

Lettuces have grown in popularity as salad ingredients as much of their inherent bitterness has been reduced by cultivation.

How the lettuce first grew

Lettuce – the world's favourite salad green, also popular in sandwiches and stir-fries – has travelled a long way since its origins, in both time and distance. Its genus is *Lactuca*, in which there are about 100 species.

The domesticated lettuce (*Lactuca sativa*) originated in the Caucasus mountain range between the Black Sea and the Caspian Sea and dispersed from there in two directions: west, through the Mediterranean and Europe to North America, and east across Asia. In China leafy lettuce arrived more than 1,900 years ago, and was then developed into stem lettuce about 900 years later.

The domesticated lettuce's closest ancestor is the prickly lettuce (*Lactuca serriola*), but its relatives the least lettuce (*Lactuca saligna*) and the great lettuce (*Lactuca virosa*) have introduced some of their attributes to the modern form. Wild lettuces flower readily and their yellow flowers, like dandelions, produce fluffy, wind-dispersed seeds. By contrast, modern domestic leaf lettuces have a prolonged leafy stage. They are slow to flower, have reduced stem latex and bitterness of leaves, non-shattering seed pods and no prickles. They have inherited a wide variety of leaf forms and colours from the diversity of wild forms in the Mediterranean and Western Asia. Lettuce is divided into six types: four are leaf lettuces – butterhead, crisp head, looseleaf and romaine – the other two are stem and oilseed, which are grown for their seeds rather than their leaves.

Foragers most likely wild-harvested lettuce more than 6,000 years ago. Early clay tablets show that lettuce was grown by the Sumerians in Mesopotamia, the land between the Tigris and Euphrates rivers. From there, merchants took it to Egypt, where temple reliefs carved around 4,600 years ago depict long-stemmed lettuces with many triangular leaves. This variety, which had a reputation as an aphrodisiac, was still being grown in Egypt in the 1960s.

Stem lettuce was grown as an oil crop (the oil is still available for its food and cosmetic health benefits), while leafier forms were selected by the ancient Greeks and Romans after they conquered Egypt. The Romans served lettuce after meals with a salty dressing – hence the English word 'salad', derived from the Latin *sal* for salt.

Soft-leafed lettuces with long straight leaves were named cos, possibly because of horticulture production on the Greek island of Kos. They were later named romaine lettuce by the French because the Romans traded them. Butterhead, with a round head of soft leaves developed in Europe, was less bitter than cos.

Cos or romaine lettuce is today the most widely eaten lettuce in the world.

In Europe many varieties emerged later, among them the crisp, round-headed Batavia (*Lactuca virosa*), which was developed in France and nicknamed 'Le crunch'.

Lettuce was taken to the Americas in the 1490s, and subsequently soft-leafed and crisp forms became the most popular. The problem was that lettuce could not be transported long distances to market, but in 1894 the Burpee seed company developed robust varieties from the Batavia lettuce that could be transported by rail from California – the USA's major producer – to east coast cities. These long-distance travellers got the name iceberg lettuces because they were kept fresh by crushed ice. The most popular varieties – New York and later Great Lakes – predominated in all forms of catering in the USA in the mid-20th century. Although they are easy to grow, they are less nutritious than many other forms of lettuce.

How the alfalfa was made

The dawn of agriculture in western Asia around 10,500 years ago was a turning-point in human history, leading to the permanent settlements and technologies that are the basis of modern society.

It was in this region, roughly encompassing present-day Turkmenistan, Iran, Turkey and the Caucasus, that previously nomadic peoples first decided that, instead of feeding their cows, sheep, goats, pigs and horses on the wild plants and other vegetation that they happened upon during their wanderings, they would grow crops for animal feed in settled locations. Among the plants they chose for this purpose were clover, vetch and, above all, alfalfa (also known as lucerne). The Medes, an ancient warlike people of what is now Iran, grew alfalfa as horse feed and also used it for their own food and medicinal purposes.

From there, the plant spread throughout the Middle East, and by 3,000 years ago it was established as animal fodder throughout most of Eurasia, from Portugal to China. (It is noteworthy that

The main use of alfalfa has always been as animal feed, but it also makes an attractive garden flower.

the crop spread fastest along similar latitudes: it was slower to take hold longitudinally above and below the Equator because doing that would have required adjustment to changing day lengths.) In Europe, cultivation of alfalfa then declined in the Middle Ages, but it was reintroduced by the Moors from North Africa to Spain in the 15th century, and from there it crossed the Pyrenees and regained its former prominence across the greater part of the whole European landmass.

At around the same time, Spanish conquistadors introduced alfalfa to South America. The crop took a while to become established in North America – alfalfa seed was introduced into California from Chile in 1850 at the time of the Gold Rush, the beginning of a rapid and extensive introduction of the crop throughout the western USA to feed horses and cattle. It was later introduced to Australia and New Zealand. Today, alfalfa is grown all over the world, with major producers including the USA, China and Canada. In the USA it is now grown on around 9 million hectares (23 million acres of land) and is the nation's fourth largest acreage crop.

What makes alfalfa so special as a fodder crop? It is a legume, so it has bacteria in its root nodules that fix atmospheric nitrogen to feed the crop and benefit any subsequent crops. Growing to heights of up to 1.2 m (4 ft), it is taller than clover, and thus more productive per plant. It also has deep roots, is very drought-resistant and long-lived. All these properties make it valuable for regenerative and organic farming production. Although the market for organic fodder production compared to conventional production is tiny, it is growing. Alfalfa can be cut as hay or haylage (bailed and wrapped in plastic or preserved using fermentation) to feed livestock. From an ecological standpoint, the flowers attract a variety of bee species which can benefit surrounding crops.

Alfalfa is nutrient- and vitamin-rich, with potential antioxidant, anti-inflammatory and cholesterol-lowering effects. Sprouted seeds are a health food eaten as a herbal supplement. The plant is also a rich source of dietary fibre for livestock feed.

There are many alfalfa varieties adapted to a range of climatic conditions, ranging from eco-types with winter dormancy for northern USA and Canada to varieties for southern USA, where mild winters allow them to grow throughout the year. There are two cultivated alfalfa species: *Medicago sativa*, with blue/purple flowers, and *Medicago falcata*, a shorter, hardier and drought-resistant variant with yellow flowers. A complex of varieties has also been created from interspecific hybridization. Most recently, alfalfa has been genetically modified, leading to environmentalist concerns about the potential dangers of cross-pollination and contamination of natural organic alfalfa.

The various apple species have many differences but what they all have in common is lovely blossom.

The crab-apple that played with the bee

Blossom time heralds a fresh start: the arrival of spring after the dark, cold days of winter. Crab-apple species and hybrids are among the earliest of the year's apples and can have long flowering periods.

Originating in central Asia, then brought to Europe along the Silk Route, crab-apples evolved on the journey and became domesticated. Some are grown for the beauty of their flower displays or tree forms, others to make crab-apple jelly. They also have a role to play as pollinators of commercial apple varieties.

The term 'crab-apple' may refer to any of more than 50 wild *Malus* species and named hybrid varieties. Many crab-apples cross-pollinate with domestic apples. The domestic *Malus domestica* is itself a hybrid of four crab-apple species: mainly *Malus sieversii*, which has large fruit-like domestic apples, together with *Malus baccata* (the Siberian crab-apple), *Malus orientalis* (the eastern crab-apple) and *Malus sylvestris* (the European crab-apple).

The edible apple is a hybrid of four crab-apple species.

The Tian Shan mountains in central Asia have fruit forests, which are a valuable resource of diversity. Thanks to random pollination by bees, these forests still contain wild ancestors of domestic apples and a wide range of flavours, from sour to sweet, including hazelnut, aniseed, liquorice and berry. However, they are under threat of destruction from human development, as are other wild crab-apple species globally. Human encroachment is also leading to the dilution of wild populations of crab-apple species as bees cross-pollinate them from nearby domestic apple orchards. Some of the 'wild' European crab-apples are hybrids: they can be identified by trees that have pinkish flowers rather than the white flowers that are typical of the wild species.

Some crab-apple fruits are sold in various parts of the world. Among the most noteworthy is the Taiwan crab-apple (*Malus doumeri*), which appears in season in street markets in Asia. Crab-apples are remarkably resistant to disease, and they are used in breeding programmes with domesticated apples – where humans take over the role of bees in pollination! In the USA there is interest in using North American crab-apple species, such as the Oregon crab-apple (*Malus fusca*), in breeding programmes adapted to local conditions for hard cider-making.

Bees and other insects are needed to pollinate the fruit-tree flowers to ensure a good yield and prevent smaller and misshapen fruit. Domestic apples have staggered flowering times: early varieties come into leaf and flower before later varieties awaken from their winter slumbers. Most domestic apples normally don't self-pollinate. Varieties are planted in cross-pollination compatibility groups to maximize successful pollination. Crab-apple trees are commonly planted within orchard rows or grafted on top of domestic apple trees because they provide a source of flowers over a prolonged period for bees to cross-pollinate with domestic apple trees.

Honey-bees are brought into orchards in large numbers to ensure saturation of the area with bees and hence a good fruit set. Other bees and insects also play significant roles in cross-pollination: wild bumble-bees, solitary bees living in holes in the orchard floor, hoverflies and flies are some of the unsung heroes of pollinating fruit trees.

The catmint that walked by itself

The smell of catmint has a famously stimulating effect on felines. Humans drink catmint tea to help them relax. Other creatures dislike and shun the plant. It's spread throughout the world, but the question is, is it a colonizer or an invader?

There are approximately 250 perennial and annual catmint species in Europe, Asia and Africa – they are native to all three continents – and they have naturalized in North America and New Zealand. The plant's botanical name, *Nepeta*, is derived from Nepi, a town around 50 km (30 miles) north of Rome, Italy, from which it either originated or (perhaps more plausibly) with which it was most associated by the ancients. The leaves and square stems of catmint are reminiscent of nettles. Most cats like to roll in the foliage of *Nepeta cataria* and related species because they give off an aroma of nepetalactone, a volatile organic compound that gives the animals a temporary sense of elation. Some people plant catmint near their houses to attract cats in to control rodents; others plant it away from their houses to keep cats at a distance!

In general, the best-looking catmint flowers are the least attractive to domestic felines.

Catnip oil attracts lacewings, which eat aphids and mites. Although bees and butterflies are attracted to pollinate the flowers, many other creatures, including deer, rabbits, flies, mosquitoes and cockroaches, hate the smell of catmint and avoid it. Planting catmint around the house also prevents termites invading the woodwork – their favourite food! Catmint extracts can be applied to the skin as an insect repellent.

There are many garden-worthy varieties of catmint: some are quite compact; others can form clumps up to 1 m (3 ft) wide. Catmints flower for about two months between spring and autumn. It is a good idea to cut the flower stems back after flowering to stop them running to seed and self-seeding around the garden. This will also make the plants more compact and encourage them to flower again. Catmint plants are common garden escapees that will colonize new areas: seed-eating birds distribute the seed heads, and humans can inadvertently move them into garden waste or soil.

Let's consider what happens to some *Nepeta cataria* seeds beyond the garden gate. If land has recently been cleared for farming, industry, or by forest fire, it appears initially devoid of life, but soon plants brought in by water, wind, birds and animals start to recolonize the area. Then, as the land becomes more fertile, other species come in. These new arrivals all benefit from the early colonizers, and some may outcompete them. Eventually stable mixed communities, such as forests or steppe grasslands with long-lived species, can form. These so-called 'climax vegetations' remain stable unless any of the variables (climate; human activity; animal population) change.

Nepeta seeds are dropped by birds and germinate on fallow land. They have many of the qualities of pioneer and second-generation colonizing plants. They grow quickly: some can flower in their first year from seed. They can spread to form local colonies on farmland from new plants developing from rhizomes. They tolerate poor soil and can adapt to all but badly drained soil. They prefer full sun, but can tolerate shade. They are most commonly found in old fields, open woodlands, fence lines along the sides of railways, and on riverbanks. They are largely untroubled by herbivores.

So did catmint invade North America and New Zealand or did it colonize those landmasses? Insofar as the plant contains biochemical compounds not present in the native flora and lacks natural predators, it can be regarded as an invasive species. Moreover, it's often classified as a weed, and hence undesirable by definition. On the other hand, invasive species have the potential to cause significant ecological and economic damage. Catmint does no such thing: it coexists with native species, and can hence best be described as a local colonizer.

The versatile buttercup is able to spread both sexually and vegetatively.

The buttercup that strayed

Many plant species propagate themselves vegetatively rather than sexually, using side-shoot suckers, runners and stolons (stems running over the ground), or by offsets from their storage organs (bulbs, rhizomes, corms and tubers).

The vegetative (asexual) method of plant reproduction has both advantages and disadvantages compared to sexual reproduction (via pollination through bees and other insects). Asexual propagation can be rapid, which increases the ability of plants to colonize new areas. It also avoids problems associated with pollination, seed development and germination. On the other hand, it can lead to a reduction of diversity, which makes plants more susceptible to pests and diseases, although vegetative propagation can generate new variants through mutations, which can be improvements on the original clone. There again, the production of genetically identical individuals can be very useful in agriculture, horticulture and landscaping, where having identical plants has obvious benefits. For example, domesticated bananas are clonally propagated from shoots.

The buttercup (*Ranunculus*) is abundant and widespread throughout Europe and North and South America. There are more than 1,700 species, and they may spread either sexually, or vegetatively via stolons. Their petals are often smooth and shiny, especially in the yellow varieties, creating a mirror-like reflection designed to attract pollinating insects and regulate the plants' own temperature.

The buttercup is highly versatile, but nevertheless not all species are flourishing. For example, the Illyrian buttercup (*Ranunculus illyricus*) is a critically endangered species found in dry, natural and semi-natural grassland plant communities in Europe and Asia. It is under attack from three angles: loss of habitat as crop lands are increased; the use of chemical fertilizers; and competition from invasive species. Researchers in Poland are studying its methods of reproduction in an effort to understand how best to increase its population in its natural habitat.

The Illyrian buttercup sends out stolons that produce new plants in the first year, which in turn produce rhizomes to store food during the dry summer months. Although the species flowers and produces viable pollen, seed production in trials was low: this may be because of a shortage of pollinating insects, but is more likely due to a lack of diversity in the population, as the Illyrian buttercup doesn't readily self-pollinate. So a greater diversity of seed-raised plants could increase seed production.

Another problem is seed dormancy, because seed germination requires temperatures below 10° C (50° F) between autumn and spring for germination to be successful. Following germination, young seedlings suffer from grassland competition, especially from invasive robust grass species. The solution is to graze or cut the vegetation to allow young seedlings to grow and thus restore the natural grassland habitat.

Sexual reproduction can be prolific if the plants are allowed to establish themselves – indeed, it can be much more productive than vegetative reproduction, which also has a greater cost for the parent plant. However, vegetative propagation is 100 per cent successful in forming new plants. In the Polish research project, vegetatively produced plants flowered in the first year, whereas it took about three years for seed-raised plants to reach flowering size. So in this case natural vegetative propagation has an advantage in producing large plants that flower rapidly. From a conservation standpoint, reintroducing seed-grown plants would have an advantage in conjunction with grassland management to increase natural species diversity. The message here is that both sexual and vegetative propagations have a place in maintaining diverse communities of plants.

How the rose got its thorns

Ouch, that hurt! We often experience sudden, sharp pain inflicted by thorns, spines and prickles on plants when we cut a bloom or pick a fruit. These sharply pointed structures are the plant's defences to deter animals from eating them.

But next time you remove a rose thorn from your skin, consider this: roses have no thorns – they have prickles! Unlike thorns (which are derived from shoots) and spines (which are derived from leaves), prickles develop from just below the outer layer of the stem (the epidermis) and have no deep connection to the plant's central plumbing system (which transports water and sugars up and down the stem). So they break off the stem easily, which thorns and spines do not.

There are around 200 species of rose, and analysis of a wide range of them has identified two categories: glandular (which develop glands on some of their prickles) and non-glandular. The majority have smooth prickles, for example *Rosa omeiensis* has spectacular, large translucent prickles on young branches,

Roses without prickles are very rare in nature, and are most likely products of selective breeding by humans.

but a few species, such as *Rosa bracteata*, have prickles with stems covered in hairs. Glandular prickles are less common than non-glandular forms. Some glandular thorns are smooth, some hairy, others branched – notably those found on the flower buds of moss roses adding to the flower's fragrance.

Some roses have no prickles – these include a few climbers such as forms of *Rosa banksia* and the climbing hybrid tea rose Zéphirine Drouhin, whose lack proves the undoing of a suspect in the Agatha Christie novel *Sad Cypress*. Prickles are problematic for the cut flower industry and have to be removed, but in nature roses without defensive prickles are very rare and are most likely a result of human selection.

These physical defence mechanisms, which prevent animals from eating them, are present in a vaste range of plant species. Many citrus trees, such as the Lisbon lemon and the West Indian lime, have thorns, and thorns are also found on some tubers of the African bush yam (*Dioscorea praehensilis*).

Spines – firm, slender, sharp-pointed structures – have vascular connections, just as thorns do. They can be derived from leaf stalks (petioles), such as in the North American semi-succulent ocotillo (*Fouquieria splendens*); from leaflets, such as in date palms; or from stipules (leaf-like appendages at the base of the leaf stalk), present in some euphorbias.

Spines also offer a shade mechanism for some plants such as the saguaro cactus (*Carnegiea gigantea*), an iconic plant from the Sonoran Desert, Arizona, which can live for up to 200 years.

Pointed structures can also help climbing plants to scramble over obstacles. For example, the hook-shaped spines on members of the rattan family are used to help climb over other plants (rattan is unpleasant to harvest – luckily the sharp spines are removed before it is used for furniture-making!). As for human benefits, there are many – armoured plants, such as the hawthorn, can be used to protect property, crops, newly planted trees and livestock; while others can be used to produce fishing hooks and gramophone needles!

Grazing sheep in Chile can get entangled in the huge hooked spines of Puya chilensis *with often fatal consequences.*

Snowdrops have adapted remarkably to survive in very low temperatures.

The snowdrop that grew in winter

It's not only cars that need antifreeze in winter! Some plants have evolved their own coolant systems to enable them to grow, flower and set seed at low temperatures and survive freezing conditions.

The appearance of the first snowdrops brightens winter days and is a reminder that spring is just around the corner. Snowdrops make the most of the lack of competition before trees come into leaf, growing well in moist, well-drained humus-rich soil in dappled shade. Their bulbs are adapted to harsh winter conditions, storing energy and natural antifreeze which allow them to survive in temperatures as low as -4° C (25° F). The best-known snowdrop species is *Galanthus nivalis*.

Snowdrops are native to Europe and western Asia, and most species flower in late winter. As temperatures drop in winter, their root systems start to grow and absorb nutrients and water to feed their new leaves. Their tough leaf tips force their way through soil that is often semi-frozen or snow-covered.

Galanthus flowers move upwards to attract pollinators when temperatures rise above 10° C (50° F), bending their flower heads for protection when temperatures drop again. Their natural antifreeze alkaloid proteins lower the freezing point of water, preventing cell damage from ice crystals and enabling them to bounce back after being frozen. Their chemical arsenal has a range of other functions too. An agglutinin protein protects against attack by aphids and leaf-hoppers, causing sugars they suck up from the plant to form indigestible clumps in their stomachs. The alkaloid galanthamine, first found in snowdrops in the 1950s, is now used to treat Alzheimer's disease.

The genus *Helleborus* contains around 15 species, some of which can survive temperatures as low as -15° C (5° F) . One of these species, the Christmas rose (*Helleborus niger*), is native to woodland and open areas in northern Italy and the central and eastern Alps. Its leathery, dark blue-green evergreen leaves enable it to photosynthesize in low light. Its deep roots can access moisture and nutrients even when the ground is frozen or covered in snow. After its flower heads droop under the weight of snow, the stems and large white flowers recover virtually unaffected by freeze-thaw stress. Like snowdrops, hellebores are packed full of cryoprotectant chemicals and are toxic to animals. Their flowers are pollinated by bees flying on mild winter days, and they can self-pollinate even under snow. Snowdrop and hellebore pollen can germinate at sub-zero temperatures.

In the high Andes, around 4,000 m (13,000 ft) above sea level in Peru and Bolivia, two potato species, *Solanum juzepczukii*

Hellebores are adapted to survive in cold, hard conditions thanks to their deep roots and thick leaves.

and *Solanum curtilobum*, contain alkaloids that enable them to grow on sunny dry days and during freezing cold nights. This makes them bitter to eat and inedible, but human ingenuity has developed a way of removing the alkaloids: farmers harvest the potatoes in winter and make small piles that are exposed to freezing night temperatures and warm sunny days in order to freeze-dry them and remove water. Then workers stomp on the tubers to remove the skins and break down the cells. Once washed, the tubers are left to dry in the sun. They are then alkaloid-free and good to eat. Known as *chuños blancos*, these frozen white potatoes are added to soups and stews, and can be safely stored for several years.

How the tulip got so many colours

Beautiful flowers aren't essential for our survival, but they are for the human spirit. People have collected and grown flowers for millennia, and done much to improve the colours of many species, particularly tulips and dahlias.

Tulips grow naturally in the Middle East and central Asia. Wild plants were probably selected and cultivated in Iran in the 10th century, and from around 1055 in Turkey, where they became a symbol of the Ottoman Empire. New forms and flower colours were likely to have arisen in gardens in these countries. In the 16th century they were introduced into the West and Carolus Clusius raised their profile when he completed the first major study on tulips in 1592, noting the variations in flower colours. They were rapidly introduced into Europe and became much sought-after by the wealthy. In the early 17th century, particular varieties with contrasting light and dark-coloured 'broken' petal patterns became highly desired for their beauty. They were difficult to propagate and their rarity led to 'tulip mania' in the Netherlands, where high prices were

Tulip lovers have as many as
130 colour varieties to choose from.

paid for individual bulbs. Between 1634 and 1637 the most desirable bulbs were sold for more than the price of a mansion, but they were overvalued: the market crashed, ruining many people. We now know that the sought-after flame colour pattern was caused by tulip break virus, so most of the infected bulbs rapidly lost vigour and died out. These days, the tulip industry centred in the Netherlands actively combats viruses in tulips as they are contagious and debilitating.

Tulips are grouped in 16 divisions according to their shape, origin and flowering time. They flower in almost every colour, either a single shade or bicoloured, although pure blue pigmented flowers are not available naturally. The rarest – and hence the most valuable – shade is black. Some of the flowers that are touted as black are in fact a very deep purple, although the difference may be unclear to the human eye.

Mirabilis jalapa *likes full sunshine so flowers in mid-afternoon: hence its common name – the four o'clock flower.*

Mirabilis jalapa, commonly known as the marvel of Peru and the four o'clock flower (the hour it normally blooms), is remarkable for a number of reasons to do with its genetics. Different colours and patterns can occur in the same flower; flowers can change colour as they mature, from yellow to pink, or white to light violet; if red plants are crossed with white plants, their offspring will be pink – an exception to Mendel's law of dominance, as in this case the red and white genes are of equal strength, so neither dominates. This is known as incomplete dominance.

Dahlia species from Central America were introduced into Europe from 1798 onwards. By 1826, double varieties, mainly with purple or tinged purple flowers, were being grown almost exclusively, and there was very little interest in the single forms. Introduced from Mexico in the 1870s, *Dahlia juarezii* was an entirely different type of flower, with a rich, red colour and a high degree of doubling, with its petals rolled backwards rather than forwards. This new form has revolutionized the dahlia world. It has a complex genetic make-up, with eight sets of chromosomes from wild relatives. Several genes control individual characteristics, such as colour and flower form, so there is plenty of scope for inheritance of a diversity of genes for the expression of flower colour.

The English bluebell droops to one side, unlike its Spanish counterpart, which remains upright.

The bluebell that joined the gang

Among the most beautiful sights in Britain in early spring are the carpets of dark blue-violet flowers of the common bluebell (*Hyacinthoides nonscripta*) on the floors of still-sleeping deciduous woodlands.

Bluebell flowers are a rich source of nectar and pollen for spring bumble-bees and hoverflies. Although they grow naturally from northwestern Spain into the British Isles, their main concentration is in Britain, with up to 50 per cent of the species present there. They have also naturalized in the USA. How have they succeeded and what are the threats to them?

Large groups of common bluebells are an indicator of ancient deciduous woodland – they flourish in dappled light, where there is limited plant competition. They harness the restricted light to photosynthesize, flowering early in spring before other plants come into leaf. They benefit from shade as the over-canopy comes to life in the summer; they also thrive when growing with bracken and Japanese knotweed. There are concerns about habitat loss, and climate change is causing other plants to leaf up earlier, potentially shading out the bluebells. Woodland management to allow light into the understorey helps maintain bluebell populations. The common bluebell also grows in fields, hedgerows and coastal locations. It is tremendously successful at spreading by seeds, which germinate over winter and from bulb offsets. One reason for this success is arbuscular mycorrhiza, a symbiosis between plant roots and Glomeromycota, an ancient phylum of fungi, which allows the plants to obtain the water and nutrients they need to flourish.

Bluebells were able to form large colonies in Britain when wild boar that root forest floors died out in the 1700s. However, these animals have recently been returned to some woodlands, and they could impact on spring displays if their numbers rise.

Another threat to bluebells in Britain has been increasing demand for their bulbs. There are legal protections preventing collection of bulbs from the wild, but they can be sourced legitimately from authorized growers. There has also been concern about the risk of genetic contamination from the Spanish bluebell (*Hyacinthoides hispanicus*), an alien species that has pale blue, pink or white flowers and is grown as a garden ornamental. Fertile hybrids with the common bluebell started to appear in the wild in Britain from the 1960s. Surveys have found one in six UK broad-leaved woodlands contained the Spanish bluebell or the hybrid. There were more cases near urban centres from garden escapes. There are also risks from the Spanish or hybrid bluebells being sold as native species and being planted in the wild. Recent research has shown that

Stinging nettles are native to Eurasia but are now found in the Americas and Australasia.

although both species flower at the same time, the common bluebell is more fertile when crossed with other common bluebells, whereas abnormalities can occur when they are crossed with Spanish bluebells.

Nettles (*Urtica dioica*) are also very successful at growing in large, dark green clumps. They can adapt to a wide range of soil conditions, including very fertile farmyard or waste dump soils. They are perennial plants with deep roots to absorb soil nutrients and cope with dry conditions. They spread by seed and vigorous stems (stolons) that root along the ground. The stinging hairs on their leaves help them to ward off predators, and their height prevents them from being overshadowed, enabling them to capture uninterrupted sunlight.

How the hosta got its taste

Although plant species have developed chemical defences to ward off herbivores, insect predators, pests and diseases, some predators have evolved to overcome these defences and may even use plant toxins to protect themselves.

It may surprise gardeners nursing severely nibbled hostas back to health that the leaves of the plant contain several chemical defences that are effective against certain predators. Among these secondary compounds are bitter-tasting glycosides and saponins known for their soapy taste and foam-forming properties, which interfere with insects feeding on the leaves. Bitter- or pungent-tasting alkaloids are toxic to many herbivores, causing them neurological and gastrointestinal problems if consumed in large amounts. Hostas also contain phenolic compounds and flavonoids that have antimicrobial and antioxidant properties.

Some predators can overcome these defences. Caterpillars of the dingy skipper butterfly (*Erynnis tages*) feed on hostas

As a general rule, the bluer the hosta
and the thicker its leaves, the more bitter
and repellent it is to slugs and snails.

successfully by neutralizing the leaf toxins with enzymes in their digestive systems. Deer, by contrast, have multiple stomachs that break down and detoxify the compounds in hosta leaves. Hostas are a favourite food of slugs and snails, and the plants are at greater risk of being eaten if stressed by poor growing conditions. Blue-leafed hostas are bitter and unpalatable to slugs and snails because they have a layer of protective wax over green or variegated leaves: the thicker the layer, the bluer the leaves. Blue leaves are more pronounced in established clumps and last longer in moist but well drained shady sites. Slug-resistant varieties with thicker, ridged blue leaves include Sum and Substance, a *Hosta sieboldii* cultivar.

Asclepias or milkweeds, a family of flowering plants in the Americas and Africa, are renowned for their toxic defences. North American species grow on ranges and abandoned farms, on roadsides, in pastures and ditches and on wasteland. They are chemical fortresses that have evolved to prevent attack by most insects and herbivores. When they are damaged they produce a milky latex (hence the name milkweed). This toxin contains cardiac glycosides that interfere with the sodium-potassium pump, an enzyme found in the membrane of animal cells that enables heartbeat and nerve-firing. Sheep and cattle have been poisoned by eating milkweed. Even horses and humans can suffer cardiac arrest if they consume enough milkweed toxin.

Some insects have evolved gene mutations controlling the sodium-potassium pump mechanism to neutralize the glycosides. The best-known are monarch butterflies, whose caterpillars feed exclusively on milkweed leaves. The plant's toxin means there is a lack of competition for the leaves, so there are plenty of them to eat, and ingesting toxins from the leaves provides the butterflies with protection against predators. Their ingestion of higher levels can result in brighter orange wings: this is an example of aposematism (the use of coloration to warn off potential predators). Some parasites, birds and animals have evolved the same detoxifying mechanism so that they can eat monarch butterflies and their larvae.

Monarch butterflies undertake long migratory flights annually. In North America, populations migrate between southern Canada as far south as Florida and Mexico. These migrations are multigenerational, with monarch butterflies pupating from milkweed en route. Populations have plummeted in recent times due to habitat loss and spraying milkweed plants with herbicides. To counter this, monarch 'waystations' have been created – refugia planted with milkweed species suited to feed the insects in transit.

A monarch butterfly feeding on milkweed flowers.

Hydrangeas can bloom from mid-spring through summer to early autumn, after which they start to lose their leaves.

The hydrangea that shed its leaves

Hydrangeas have always been highly popular ornamental plants, grown for their distinctive colours and large flower head shapes. Some deciduous species also provide wonderful leaf colours in autumn before they drop.

As with most deciduous plants, the leaves of hydrangeas do not have much function in autumn and can get damaged, so it is better that they drop them and start afresh each growing season. Leaves are the primary method for many plants of intercepting the sunlight and converting it into chemical energy for growth via photosynthesis; they are supported by the vascular system for the transportation of water, nutrients and energy through the leaf stalk to the rest of the plant. Plants can also shed leaves when resources are in short supply, as in times of drought. This allows the plants to move resources around effectively to survive changing conditions: their leaves are disposable assets.

In spring, hydrangeas awaken from their winter slumber as the days lengthen, soil and air temperatures rise, and water and nutrients increase their flow. Buds emerge on their bare branches and then open to produce the annual crop of fresh green leaves that unfurl and gradually expand to their full size.

The chemical processes of the leaves are constantly changing as daylight increases into summer and then diminishes into autumn. Hydrangeas have photoreceptors that detect light: two of them – phytochromes (which respond to red light) and cryptochromes (which are sensitive to blue light) – detect varying day lengths.

The predominant pigment in green leaves is chlorophyll, which is essential for photosynthesis. As the days shorten in the second half of the year, chlorophyll production slows down and eventually stops altogether. At the same time there is also a reduction in the growth regulator plant hormone auxin, and this initiates the development of the abscission (a layer at the base of the leaf stalk). This layer gradually blocks off the flow of water to the leaves and nutrients to the rest of the plant until completely sealed off. Once that has happened, the junction between the leaf stalk and the twig is fatally weakened; finally the leaf is blown away, leaving a scar on the twig.

The time that trees and shrubs take to colour up is controlled by genes and varies from species to species. The quality of the display is modified by temperature – warm autumns can reduce the impact, cold can intensify it, while wind can blow leaves off early. Soil moisture, drought and heatwaves can also impact the display. Trees such as hickories (*Carya*) and aspen (*Populus*) with predominantly yellow leaves in autumn contain high levels of carotenoid pigment in the leaves. Hydrangeas such as *Hydrangea paniculata* and *Hydrangea quercifolia*, as well as the leaves of oak (*Quercus*) and dogwood (*Cornus*), contain high levels of anthocyanin pigments that produce autumn displays of red and brown leaves. Some cultivars have particularly good autumn leaf colours, among them Merveille Sanguine (*Hydrangea macrophylla*), a mophead hydrangea with pink/red flowers that develops spectacular autumn colour when the leaves turn a brilliant scarlet.

Hydrangeas also lose leaves at other times of the year because of the stress caused to them by getting too much or too little water. Prolonged or sudden environmental stresses, such as temperature fluctuations, strong winds, or exposure to direct sunlight, can also affect their leaves. Nutrient deficiencies – if the soil is lacking in nitrogen, phosphorus or potassium – will also inhibit or prevent healthy growth. Pests and diseases are additional potential problems – spider mites and aphids feed on leaves, while powdery mildew and bacterial leaf spot can cause the leaves to turn yellow and fall off.

The camellia that turned shiny

Camellias have long been prized for their glossy foliage and bright, fragrant flowers. In Britain for many years the species was grown only in greenhouses because no one realized that _Camellia japonica_ is hardy in the local climate.

Camellias are woodland shrubs or trees that prefer varying degrees of shade, although some taller species, such as _Camellia sasanqua,_ can tolerate more direct sunlight. Young shrubs can be difficult to establish as they are more sensitive to leaf scorch than older established plants. While establishing themselves, they benefit from shade from a wall, fence or other plants nearby. The best planting location is a northwest-facing site with some shelter behind the plant.

Camellia species are notable for their reflective mature leaves that offset the beautiful springtime blooms. The shiny surface is created by a lipid and waxy layer on the leaf surface, which changes in shininess and thickness during the year. This is the cuticle – a layer that prevents the plant losing water and its

Camellias have distinctive shiny leaves that help them retain water and stop them from getting too dry.

leaves drying out. The leaves play an important role in reflecting and scattering light. Their reflective nature may attract pollinators, particularly bees. The cuticle also protects leaf tissue against damage by ultraviolet light, hot, dry conditions, pests and diseases, pollution and harmful chemicals. The layer controls gaseous exchange through the leaf surface of oxygen and carbon dioxide for photosynthesis. It also controls water evaporation and water uptake reflecting the soil conditions.

The plant's health is indicated by the condition of its leaves. Camellias don't tolerate dry or waterlogged soil. If they are overwatered, their leaves will turn a yellow-bronze colour and become limp; if they are too dry, their leaves will become crisp. If the soil is too alkaline, the camellia can't take up iron, so the plant's leaves begin to turn yellow. This can be remedied by acidifying the soil. The presence of salts causes browning at the edge of the leaves. If you see leaves wilting despite the soil being wet, the chances are that root rot has set in, and the plant is struggling to absorb water. As a guide, if the soil is dry to a depth of 5–7.5 cm (2–3 in), the camellia should be watered. So, everything in moderation – and select a site with well-drained, acidic soil.

Camellias are also the basis of the world's most popular drink – tea. Tea is obtained from the leaves of *Camellia sinensis*, a white-flowered shrub originating in China. China tea (*Camellia sinensis* var. *sinensis*) produces small leaves and is hardier than Assam tea (*C. sinensis* var. *assamica*), which has much larger leaves. Today there is interest in growing tea in cooler climates such as that in Britain. Indeed, tea estates have been planted in sheltered locations with mild winters, such as in Cornwall. *C. sinensis* is hard to establish as it requires a very specific combination of climate conditions. Tea bushes are grown in rows and pruned to produce broad, flat-topped or domed shrubs about 90 cm (3 ft) tall to ensure the maximum surface area for new shoots to be produced. When the bushes are established, they are harvested as often as indicated by the rate of growth.

The quality of tea depends on how much of the young shoots are harvested, and this differential is reflected in the price of the tea sold. White tea (young buds before they fully open) is hand-harvested in very small amounts and is consequently the most expensive. Orange pekoe – regarded as the best of the black teas – consists of two unbroken leaves and a bud. It is also picked by hand and very pricey. Mechanically harvested tea is of lower quality as the leaves thus obtained are older and hence not as tasty.

Harvested leaves contain terpenes – volatile compounds that give the tea its distinctive floral and citrusy notes, which play a vital role in the flavour and aroma of the final product. Making black tea involves the wilting of leaves, which allows enzymes in the leaf to interact and start the fermentation process. To produce green tea, the leaves are not wilted but steamed or fried to prevent oxidation.

Passionflowers originated in the wild in the Americas but are now found in gardens throughout the world.

How the passionflower bore fruit

The pulp of the passionfruit tastes sweet as an encouragement to birds and animals to eat it. They later distribute its seeds through their excrement in a process known as endozoochory.

When birds or animals eat fruit, in some cases the digestive processes of the gut don't harm the seed. Moreover, they may speed up germination by removing inhibitors – perhaps thinning the seed coating and mobilizing nutrients in the endosperm of the seed. Birds and animals make brilliant dispersal mechanisms as seedlings develop in nutrient-rich compost – the consumers' poo – and end up being released into new, sometimes far-flung, locations.

There are more than 500 species of passionflower, and most of them are found in the tropics in South America. Maypop (*Passiflora incarnata*) is a hardy temperate species that grows as a weed in the USA. Elsewhere it is cultivated commercially for its fruit and also for its leaves, which are used for herbal tea to help people sleep. It has beautiful purple and white flowers and can be grown as an ornamental. When fully ripe in autumn, the egg-shaped fruits are yellow, a colour that attracts birds and small animals to eat the whole fruit or just the sweet pulp. These creatures then distribute the seeds either by messy eating habits or in their droppings.

These distribution methods are employed by a wide range of fruiting species, and is particularly effective where the seeds are small and can't be separated from the fruit, as is the case with strawberries, blackberries and gooseberries. Successful seed distribution can cause problems with invasive species, such as the banana passion flower (*Passiflora mollissima*). As a result of pigs eating the fruit and spreading the seeds, this non-native species has become an invasive weed in Hawaii, smothering other plants.

In some cases, a loss of habitat or of key animals or birds can lead to lack of distribution. Avocados in Latin America are a case in point, where a giant ground sloth that used to munch through avocado fruit, distributing their seeds, went extinct 10,000 years ago, and nothing replaced this except humans discarding the seeds of commercial avocados!

One of the most remarkable stories concerns the Brazil nut tree (*Bertholletia excelsa*) in the Amazon, which relies on aguti, very large rodents, for seed distribution. Brazil nuts grow on tall trees inside large, very hard pods named cocos, and only the aguti has sharp enough teeth to chip away at the pods to get the seeds out. They eat some of the Brazil nuts but hide others,

Australasian cockatoos enjoy passionfruit and distribute its seeds in their droppings.

intending to come back for them later, but they evidently have terrible memories because they often forget where they secreted their foodstores, and some of these hidden seeds germinate and grow up as replacement trees.

In a similar way, bears eating apples are believed to have played a key role in spreading apple trees from their origins in central Asia (the seeds are protected in the core of the fruit). Coffee berries are also eaten by animals that digest the sweet pulp and excrete the seeds, thus aiding their distribution. One of the most unusual – and expensive – coffees in the world is kopi luwak, which is also known as civet coffee because it is made from semi-digested beans excreted from these small, catlike mammals in Indonesia. Sadly, demand for this rare drink has resulted in civets being caught and force-fed coffee berries.

How the daffodil got its colour

The daffodil – the common English name for the genus *Narcissus* – is one of the earliest and most colourful spring blooms. Daffodils are immensely popular as cut flowers and plants.

Daffodils grow and flower rapidly from bulbs. They contain toxic alkaloids to protect themselves from being eaten, which is especially important when there is a limited range of flowering plants growing early in the spring. They can flower from the end of the year until about May, depending on the different amounts of winter cold the bulbs need to grow and flower. Their flowering time varies locally and regionally. Flowering early in the year has its merits – there are few other plants to compete with it, so the next flower a pollinator encounters is likely to be another daffodil! The disadvantage is that it may be very cold and there are not many insects around: as a consequence, daffodils remain in flower for a long time – which also makes them a great cut flower. To compensate for the lack of insects, daffodils can self-pollinate their flowers.

The yellow colour of daffodils enables them to stand out against green landscapes and thus attract pollinators in the low light of spring.

Winter and early spring flowers such as daffodils are predominantly yellow or white. These colours allow them to stand out in the low light of spring in the largely green landscape so they can attract pollinators. At this point there are some bees around, but there are fewer flowers and it is cooler, which means that flies (which lack colour vision) and just a few bees are involved in pollinating. The daffodil's yellow colour is derived from carotenoid pigments that absorb violet and blue light and reflect yellow and green light. The white flower colour comes from the absence of pigments in the petals. This allows all colours of light to be reflected equally, resulting in a white appearance. At this time of year it is much better to have light-coloured flowers that reflect as much light as possible. Brighter colours – blues and reds – start appearing in later-flowering spring species.

Many early spring flowers tend to have wide corollas (flower heads) to allow in as wide a range of insects as possible. In the case of daffodils, three groups have adapted to their pollinating insects.

The first group, including trumpet daffodils (*Narcissus pseudonarcissus*), can be pollinated by short- and long-tongued bees because these insects can climb into the flowers' trumpets and forage for pollen. In a study of the Spanish species of trumpet daffodil (*Narcissus longispathus*), larger bees had to warm up their flight muscles in cold weather before they could visit these flowers, whereas small mining bees (*Andrena bicolor*) were able to fly in temperatures below their body temperature if the sun was shining. The daffodil flower not only provided them with food but also acted as a greenhouse to warm them up. The inside of the flower around the anthers (male reproductive parts) was 8° C (14° F) above the outside temperature. The bees were able to bask in the sun's warmth reflected by the flower before flying on to the next one.

The second group, of which the paperwhite narcissus (*Narcissus papyraceus*) is a leading example, is often highly perfumed, with a short, cup-shaped trumpet surrounding the floral tube containing the reproductive organs. The strong scent attracts pollinating insects such as hawk moths, long-tongued bees, butterflies and flies that feed on nectar. There are different flower forms with the stigma (female flower part) below or above the anthers, whichever arrangement is more favourable to cross-pollination.

The third group of daffodils includes the Angel's Tears daffodil (*Narcissus triandrus*). Members of this group have extended trumpets and narrow floral tubes catering to long-tongued solitary bees feeding on both pollen and nectar. There are different flower forms, rather like in primroses, with anthers either higher or lower than the stigma and style to aid cross-pollination.

Honeysuckles come in a range of colours and many of the species have a distinctive sweet smell caused by phytochemicals.

How the honeysuckle got its scent

In many gardens, flowers are planted specifically for their perfumes. Daytime-pollinated flowers generally have a sweet and fruity aroma, while night-time flowers can produce strong, sweetly floral but musky scents.

Honeysuckles – plants of the *Lonicera* genus – are a range of deciduous and evergreen shrubs and climbing species. Honeysuckle derives its name from the sweet nectar its flowers produce. To aid pollination, many species emit a strong perfume as dusk falls, but they also can be pollinated during the daytime. They have long, narrow tubular flowers in pairs, which restrict pollination to long-tongued moths at night, and bees, butterflies and hummingbird hawk moths during the day. Their flower colours vary from white, yellow and pink, which are more visible under moonlight conditions for nocturnal pollination, to bright oranges or reds, which are more visible in daylight.

Hawk moths have been identified as important pollinators of *Lonicera* species. Japanese honeysuckle (*Lonicera japonica*) has fragrant white flowers in the summer that open at dusk to encourage night-time hawk moths: these pollinators consume only the flowers' nectar, whereas bee pollinators in the daytime eat some of the pollen as well. Moth populations are in decline at a time when their significance as plant pollinators is becoming clearer: pollination by moths has a genetic advantage for the plants as the insects' range is much wider than that of bees, spreading the plants' genes further afield when they pollinate other plants. Hawk moths can detect the scent of the common honeysuckle (*Lonicera periclymenum*) from up to 1.5 km (1 mile) away. Hummingbird hawk moths, which fly in the daytime from dawn to dusk, are more important pollinators of the Etruscan honeysuckle (*Lonicera etrusca*) than bees.

In North America the orange-red flowers of coral honeysuckle (*Lonicera sempervirens*) attract daytime pollinators, including

A hummingbird hawk moth hovers by a flower and uses its long proboscis to feed on the nectar.

hummingbird hawk moths as well as bees and butterflies. In winter and early spring, honeysuckle species such as winter honeysuckle (*Lonicera fragrantissima*) and blue honeysuckle (*Lonicera caerulea*) are to be found with pale flower colours. Both have white flowers, but the latter produces dark blue honeyberries, which are edible. These species are pollinated effectively at low temperatures by bumble-bees and other daytime pollinators.

Some plant species are pollinated only at night because their flowers open at dusk and close at dawn. The queen of the night (*Selenicereus grandiflorus*), an epiphytic cactus from Latin America, takes this a step further. It has beautiful white flowers with a strong, sweetly floral, musky perfume and which open on one day only in late spring or early summer and attract night-flying moths. The large flowers have long floral tubes that make it difficult for any other insects to pollinate them. They wither by the first light of dawn and it will be another year before the plant flowers again!

Very occasionally, night-flowering plants have one specific pollinator. The soapweed yucca (*Yucca glauca*), which grows in the centre of the USA, attracts the yucca moth to pollinate its perfumed flowers. The yucca relies on the moth and the moth relies on the yucca for survival.

The curry plant that smelled like food

The leaves of many plants produce strong scents. They do this for one or more of three reasons: to attract pollinators; to warn away pests and herbivores; and to communicate with other plants.

Many of the leaves and oils derived from fragrant plants are used by humans for a variety of purposes – in cooking (lemon grass), as perfume (narcissus) and as biochemical feedstock. They are also sources of insecticides.

Sometimes very similar smells are produced by unrelated plants in different parts of the world. For example, aromas associated with curries. The curry tree (*Murraya koenigii*) found in tropical and subtropical areas of the Indian subcontinent produces curry leaf. This is an important flavouring ingredient in Indian and Southeast Asian cooking. The smell is strongest in fresh leaves when cooked with vegetables. The main leaf compound is beta-caryophyllene, which produces a mixed aroma – pungent, sweet, slightly citrus and spicy. Its leaves are a rich source of

The strong-smelling curry plant is grown mainly as an ornamental rather than for flavouring curries.

carotenoids, beta-carotene, calcium and iron. They contain anti-inflammatory and antioxidant properties and are used in Ayurvedic and Siddha medicine, while oil extracted from the leaves is used to scent soap. In nature, components of the essential oils disrupt the nervous systems of the pest insects mosquitoes, houseflies and beetles.

Aromatic compounds are reflective of the plant's ecology. For example, many aromatic plants grow in warm, temperate Mediterranean climates. Plants cannot grow very much during the dry summer; they are still photosynthesizing, but part of their strategy is to defend what they have by producing aromatic compounds from their plant tissues. For example, the curry plant (*Helichrysum italicum*) is short and woody. It grows in stony, sandy places around the Mediterranean and produces yellow, daisy-like flowers, which can be dried for everlasting displays and are used in traditional medicine. The glandular hairs on its leaf surfaces produce essential oils containing a range of beneficial chemical compounds, including neryl acetate, italidione and beta-diketones.

Although the smell of the curry plant resembles that of curry dishes, this is not what humans use it for. Instead, young sprigs are used to infuse a resinous taste into stews. Its sweet, spicy floral aroma is used in perfumery and aromatherapy. The plant's essential oils are believed to have anti-inflammatory, pain-relieving and antimicrobial effects, and have been used in traditional medicine. They have also been shown to have insecticidal properties against a range of pests, including cabbage loopers, red flour beetles and mosquitoes.

Some aromatic compounds can reflect a plant's health status or stage in its growth cycle. The katsura tree (*Cercidiphyllum japonicum*), for example, produces a strong a smell of burnt sugar when grown in a sunny position, but this scent is only perceptible in autumn, when its leaves turn yellow and start to break down, releasing into the air molecules of maltol, a natural flavour-enhancer. This conjures up an aroma reminiscent of toffee and candy floss. The powerful smell of burnt sugar some distance from the tree stops passers-by in their tracks and they may wonder where the aroma comes from.

The growth of pansies is determined by interaction between various different types of plant hormone.

What the pansy shoot said to the pansy root

Garden pansies are among the world's best-selling cool season bedding plants. They bring colour to gardens and parks through autumn, winter and early spring. They love sunny sites but can cope with shade, and even flower in snow.

The garden pansy (*Viola* x *wittrockiana*) originated in the early 19th century from the hybridization of different forms of heartsease (*Viola tricolor*) and later the mountain pansy (*Viola lutea*) and the Altai pansy (*Viola altaica*). They were originally developed as show pansies. Their characteristic dark petal blotches came from a chance mutation and resulted in the release of the first variety with a 'face' in 1839. By the 1860s, fancy pansies with large flowers and faces became popular. Today some varieties are over a hundred years old and need rigorous reselection to maintain uniformity.

Introduced in 1979, Universal pansy hybrids were a major development: they were more resistant to cold and easier to grow in winter flower beds and containers. They were a result of crossing the few plants able to flower under low temperatures. Winter pansies can survive snow, but low temperatures of -12° C (10° F) can damage flower buds, and they can be killed by very low temperatures of -17° C (0° F) when the plants roots are not able to absorb water from frozen soil.

A particular pansy type, Panola, a pansy/viola hybrid, has been bred specially to increase the species' resistance to low temperatures. It combines the cold hardiness of violas with pansy faces. Like humans, plants have chemical substances – hormones – that control growth and development in cells. Generally, cold weather protection triggers plant hormones to produce signals resulting in cold response genes triggering biochemical changes that protect plant cells. Pansies that have been adapted to cooler conditions stop flowering when the weather warms up.

Breeders are also increasing the heat tolerance of pansies so that the plants can survive hot weather after planting in autumn and be ready to flower when temperatures rise again in spring.

Pansies' growth and development are controlled by the interaction of five types of growth regulators (or plant hormones): auxin, gibberellin, cytokinin, abscisic acid and ethylene. The impact of these growth regulators can be affected by the plants' growing conditions, their genetic make-up and the interaction of the regulators. Auxin plays an important role in plant growth and is produced mainly in shoots, young leaves and seeds. Gibberellin promotes shoot growth and elongation via cell division, resulting in taller plants and delayed senescence. (During commercial pansy plant production, anti-gibberellin sprays have been used to prevent stem 'stretch', producing compact plants that are easier to establish when planted.) Gibberellin also promotes leaf expansion for photosynthesis and seed germination.

Cytokinin is thought to be involved in the response to root restrictions in the growing media or interaction with the roots of other plants. Signals from the plant's roots lead to stunted growth, which can't be changed, affecting plant quality and longevity. Abscisic acid inhibits growth and promotes dormancy in buds and seeds. It regulates stomatal closure and water uptake during drought stress: roots sense when water is in short supply, and communicate the need to close the leaf stomata (pores) by altering the concentration of abscisic acid in the shoots. Ethylene regulates a range of actions, including shoot growth, flower senescence and fruit ripening.

The balance between these growth regulators is critical in determining overall growth and development. Excess auxin can result in elongated, spindly shoots; too little results in stunted roots. Too much cytokinin can result in excessive branching and reduced root growth, while too little can result in poor shoot development. Researchers continue to improve garden pansies so that we can all enjoy these extraordinary flowers.

How the musk lost its scent

The presence of perfume in flowers can add so much extra meaning for those growing them or receiving them as a gift – a lovely fragrance is good for humans both psychologically and physically.

Flower perfume is composed of all the volatile organic compounds (VOCs), or aroma compounds, emitted by petals. In nature it serves a different purpose: to attract pollinators to cross-pollinate flowers. Modern understanding of plant genetics is helping to breed flowers that look good, last well and have beautiful aromas. In the past, however, breeders inadvertently eradicated some of their plants' most desirable properties.

The most notorious case is that of the yellow-flowered musk (*Mimulus moschatus*), which was recorded as losing its aroma in different parts of the world between 1909 and 1916. It was discovered in 1826 by David Douglas in British Columbia in 1826. Seed was sent to England, where John Lindley described and named the new seedlings, noting an 'extremely strong

The yellow musk when first grown from seed in culitvation was strongly scented, but its modern descendants are odourless.

sweet musky scent'. It became a very popular plant, grown for its perfume, in Britain, Europe and as far away as New Zealand. From the 1880s to 1909 there was a gradual loss of the musk smell, and this appeared to be the case also with wild populations, although people occasionally found wild specimens with strong aromas. There was much speculation about how it lost its smell, ranging from plant infections to the negative impact of radio waves!

A more likely explanation is that the seed grown in England had a rare recessive perfume that was perpetuated by vegetative propagation. Increasing demand then led to production of plants from seeds that were not perfumed. Another suggestion is that a popular bedding variety created by crossing it with *Mimulus luteus* lacking aroma may have diluted and eventually led to the loss of the perfume. For now we can only speculate. Based on contemporary research with other musk species, perhaps the variation in aroma observed in *Mimulus moschatus* could be linked to diversification and adaptation.

There are parallels between food and flower crop breeding programmes, where the desire to make the food look good outweighs any need to make it taste and smell good. For example, decades of breeding tomatoes for uniform colour robbed them of interesting tomato flavour genes. The quest for perfect cut flowers that travel well and have a long shelf life has led to the selection of many beautiful blooms with no perfume.

Research into the genetics of perfume expression in test species such as petunias has indicated that it may be possible to restore the biosynthesis pathways in flowers that used to smell but have since lost their fragrance – so there may still be hope for the yellow-flowered musk!

Roses have always been prized for their perfume, of which there are thought to be around 24 variations: the main ones are rose itself, nasturtium, orris, violet, apple, lemon and clover. The intensity of these aromas is generally affected by flower colour, petal texture and the thickness and quantity of the petals. Red and pink roses tend to emit the classic rose smell; yellow and white flowers typically give off more of an orris (*Iris germanica*) or nasturtium perfume.

It has recently been discovered that the biosynthesis pathway in roses differs from that of most other flowers because it depends on a special contribution by the nudix hydrolase group of enzymes which, although almost ubiquitous in nature and often regarded as no more than non-specific 'house-cleaning' fragrances, in roses generate geraniol, an alcohol that creates the inimitably sweet floral rose smell.

As understanding of the genetics of rose perfume increases, it may be possible to breed more roses that are dependably well perfumed.

Ferns emerge from the ground as fiddle-heads – tightly coiled spirals – to protect the plants as they grow.

The fern that curled up

Humans have copied the movement of plants and animals for millennia. In some cases, however, plants and humans have solved similar problems independently.

Ferns produce their spores in sporangia – hollow, walled structures or sacs – then eject them forcibly with catapult-like flicks that send them off to land in other locations and, with luck, take root there.

An agglomeration of sporangia is known as a sorus (plural: sori), and clusters of them may be seen on the undersides of fern fronds. When the time comes for the sporangia to leave the mother plant, they are exposed to the air, thus allowing water to evaporate through the thin outer walls of the cells that surround them. Increasing water tension in the cells causes cavitation – the creation of bubbles. These small air locks interrupt what would otherwise be an unbroken column of water and create elastic tension, the energy of which is quickly released, leading to fast closure and ejection of the spores as if they were being

113

fired from a catapult. Some of the spores launched in this way can catch fast air currents and be carried vast distances. This is one of the reasons why ferns are among the oldest life forms on Earth, dating back around 300 million years.

Another example of this ejection device may be seen in *Ecballium elaterium*. This is no ordinary cucumber, but rather the squirting cucumber, a native of Europe, North Africa and temperate parts of Asia, which is regarded as an invasive species. It has small, slightly hairy fruits that act as liquid-filled pressure vessels. If the fruits are tapped gently when ripe, they explode immediately, ejecting liquid and seeds at great speed. The squirting cucumber is definitely not edible, as its fruit is harmful and its juice an irritant.

The Himalayan balsam (*Impatiens glandulifera*) has a similar mechanism. Again it is easy to trigger when ripe – a gentle touch results in an explosion of seed capsules, which disperse typically over distances of up to 7 m (23 ft). By these means the species has successfully colonized waste land and riversides in a wide range of locations where it is not wanted. Like the squirting cucumber, it is highly invasive, and it has been banned by the European Union since 2017.

On a smaller scale, you can observe in your garden the seed capsules of pansies and violets breaking open and curling back before springing forward to disperse their seeds.

The pumpkin-shaped dry fruit of the sandbox tree (*Hura crepitans*) from the tropics of Latin America explodes when ripe, projecting seeds at more than 250 kmh (150 mph) over distances as great as 100 m (330 ft). This amazing expulsive force has earned the plant the nickname 'the dynamite tree'. Both the fruit and the seeds of this plant are toxic – its sap has been applied to the tips of poison arrows.

Dodders (Cuscuta), of which there are more than 200 species, are yellow, orange or red parasitic plants related to the morning glory (Convolvulaceae) family found in grassland. When they germinate, they mount the roots or stems of their host plants, sucking up their water and nutrients through haustoria – highly modified stems or roots of their own.

Another plant that propels its seeds into the air is *Cardamine parviflora*, commonly known as small-flowered bittercress or sand bittercress. It is a flowering plant of the Brassicaceae family. Ballistics studies have found that although its ejections are not always effective – many of the seeds merely fall straight to the ground – it is efficient enough to propagate itself. What is most remarkable about it is its speed: each flick takes 5 milliseconds, faster than the blink of an eye.

Germinating seedlings of the purple-flowered witchweed (*Striga hermonthica*) are chemically attracted to move towards and parasitize the roots of maize plants, sapping their nutrients and strength. They can devastate yields in Africa. Interplanting maize with the silverleaf desmodium (*Desmodium uncinatum*) legume inhibits *Striga* seed germination and ensures the maize crop yield.

How the acacia tricked the zebra

Can plants talk to each other? Can they send out alarm bells when they are under threat? Can they contact buddies to come to their aid? Perhaps surprisingly, the answer to all these questions is yes!

Groups of acacia trees are characteristic of the plains in Africa. Their leaves are a favourite food of the giraffe, any one of which can eat up to 29 kg (64 lb) of them daily – in fact, browsing giraffes contribute to the flat dome shapes of the trees. Acacias are also popular with other African herbivores such as gazelles, zebras and elephants. This might not be a problem if acacia trees grew quickly in fertile soil, but they do not; their growing conditions are poor, and they have had to adapt to survive.

A first line of defence among acacia trees in Africa is that they have lots of thorns on their branches: this deters some herbivores, but the giraffe has a long prehensile tongue that can negotiate around thorns to get the desirable leaves, and the animal's tough lips and palate can cope with twigs.

Acacia trees evolved long, sharp thorns in order to discourage herbivores such as zebras and elephants from eating them.

The acacia's second line of defence is its ability to produce high levels of tannins in leaves that are being nibbled – after about two minutes' browsing, levels can rise rapidly and they take 24 hours to return to normal! Tannins are present in a lot of plants to make them unpalatable to eat. In moderation, however, they are health-protecting compounds: we have them in some food and several drinks – they are what makes your teapot go brown and gives the dry sensation to hard cider before it is fully fermented.

The acacia cannot produce high levels of tannin in its leaves all the time – raising the level is a defence response. Giraffes have a discerning palate and don't like a lot of tannin in their diet, so they browse for a short time and then move on to another tree. There is more to this story than first meets the eye – an acacia being eaten by browsing animals can raise an alarm to warn its neighbours up to 50 m (160 ft) away to defend themselves by increasing their own tannin levels. The tree under attack emits jasmonic acid, which is carried on the wind to neighbouring trees, which can then increase their own tannin levels in around 30 minutes. In response to this, giraffes have developed a strategy to outwit the trees: as far as possible they graze facing the wind so that the warning signals emitted by the tree being eaten blow away from adjacent acacias.

All this was discovered by chance after large numbers of kudu antelope that were being farmed for their meat died suddenly in their enclosures. They had been fed mainly on acacia leaves, and autopsies revealed that the cause of death in every case was tannin poisoning.

Some acacias have a third line of defence: they provide homes for biting ants that defend them from browsing animals who dislike being bitten. The insects inhabit either chambers in swollen thorns or dome-shaped chambers made from branch tissues, and feed on the trees' sugary sap secretions. What is more, the acacia can distinguish between defending ants and less desirable insect species: the latter don't get the same VIP treatment. Defender ants have been effective in reducing the foraging of leaves by elephants.

Chemical signals are not limited to the plains of Africa – they are very widespread, something to ponder when you smell newly cut grass. It isn't just a smell to please us: it contains signals to summon insects to protect the plants from attack.

A giraffe can feed on the highest leaves of an acacia tree.

Garden peonies have long been popular in China, where they are traditionally taken to symbolize wealth.

The cool way peonies flower

Peonies are valued for their large, showy, opulent blooms in a range of colours. There are about 33 species, and in China they have been cultivated for 2,000 years as both medicinal and ornamental plants. They were introduced into Japan in the 10th century and to Europe in the 15th century.

Peonies are often used as cut flowers for wedding bouquets, indoor arrangements and gifts. In China they are known as 'The King of Flowers'. In Japan they are associated with romance. Today the main centre of cut peony flower production is the Netherlands.

Only certain varieties of peony with long enough vase lifes are are suitable as cut flowers. The variety Sarah Bernhardt is a leading example. The natural production season is short, but peonies are also grown with artificial chilling to induce flowering in parts of the world where they wouldn't naturally prosper. Some peonies don't flower for long, and a common problem encountered by gardeners is when they don't flower at all. This isn't the case when they are happy and undisturbed. For example, cultivars of *Paeonia lactiflora* are long-lived garden plants that flower reliably over many years.

To get the best out of a peony, the depth of planting the crown is critical: the general advice is to plant them so that the buds are 5cm (2in) below soil level in temperate climates, and a little deeper – up to 7.5cm (3in) – in colder environments. Peonies have no problem with winter chills – indeed, they need them to bud: that is one of the qualities that makes them especially popular and successful in Alaska.

The kurinji is one of the rarest – and rarest-flowering – plant species in the world.

Peonies can be artificially chilled in pots for cut-flower production and in areas with mild winters. There are variations in cold requirements, and some cultivars may be more suited to milder winter conditions – they need between 500 and 1,000 chill hours, roughly 20 to 42 days, at temperatures of 0°–4° C (32°–40° F). In tree peonies, defoliation (loss of leaves) over winter and during cold weather is important for changes in sugar assimilation and flower bud development, as they don't flower for long. Defoliation after flowering has been trialled experimentally and has resulted in tree peonies flowering for a second time later in the same year.

While peonies have exacting requirements, they are rewarding with their flowers. Some plants, however, flower only very rarely. An extreme example is the kurinji (*Strobilanthes kunthiana*), which blooms every 12 years, covering some of the hillsides of Kerala, India, with a lovely violet blanket that attracts visitors from all over the world.

How the chrysanthemum became a year-round flower

As members of the aster family (Asteraceae), chrysanthemums flower naturally only when the days are short in autumn. However, demand for them in florists is year-round, and necessity is the mother of invention.

Ornamental chrysanthemums vary in hardiness and flowering time. They are short-day plants, which means they flower in autumn when the nights are starting to draw in. So how come they are available in florists at all times of the year? One way is through importation: thanks to jet aircraft, chrysanthemums may be flown in from places that have the right seasonal day length to countries that do not. Another way is by artificially manipulating the duration of light.

Photoperiodism (the physiological reaction of organisms to the length of night or a dark period) influences floral induction and flowering rate in many plant species, from algae to angiosperms. (The same tendency may also be observed in many mammals, from microscopic rotifers to rodents.) In some plant species the

In nature, chrysanthemums are autumnal, short-day plants. Researchers seek ways to produce varieties that flower naturally under longer day conditions.

flowering process is unaffected by day length: such plants are classified as 'day neutral'. Other species flower only in the long days of summer away from the Equator.

Chrysanthemums and asters are among the plant species that flower only under short-day conditions, but day length – or rather the length of night – can be manipulated to induce chrysanthemums to flower outside their normal flowering season, and this enables some growers to provide year-round production in greenhouses.

Chrysanthemums remain vegetative (non-flowering) when maintained in 14 hours or more daylength, which can be created by artificial illumination. The number of long days increases the stem length of the plants until the desired length for marketing is reached. The stem apex can be removed to allow flower stems to branch to meet a desired date of bud initiation. Chrysanthemums require 14 hours or less of daylight for flower bud initiation, and 13 hours or less for flower bud development.

So what causes this effect? Light is made up of a spectrum of colours with different wavelengths. Red light (photons with wavelengths of 660 nanometres) and far-red light (730 nanometres) – the latter which humans can hardly see because it is at the end of our eyes' visual sensitivity to light quality – are influential in plant development and flowering via a phytochrome, a chemical in the leaves. This exists in two forms, one of which responds to red light, the other to far-red light: the former promotes flower formation; the latter inhibits it.

Applying varying quantities and wavelengths of light under artifical conditions has been proved to work on chrysanthemums. Using blackouts of dark cloths to cover the plants and mimic long nights is effective, but if one leaf is exposed to light or if a shaded plant is grafted to a plant exposed to light, it will flower. A ten-second flash of red light onto chrysanthemums growing under short-day conditions with long nights is sufficient to interrupt flower development, while a similar flash of far-red light doesn't affect flower formation. Understanding the plants' interaction with light helps to produce a programmable and repeatable growing system for year-round production.

The modern sweet pea was initially developed by amateur growers and} gardeners in the late 19th century.

How the sweet pea learned to climb

What would you recommend if someone is stuck at the bottom of a dark hole but can see the light above showing the way out? You might suggest they reach up and find any handhold to grasp and haul themselves up to reach the next handhold. In a way, climbing plants with tendrils do something similar to reach the light.

Climbing plants use a range of adapted organs, leaves, stems and roots to grasp surfaces. Those using tendrils or modified leaves to climb up are termed 'vines'; if they wrap their stems around an object, they are referred to as 'bines'. Climbers can be annuals, such as sweet peas, or perennials such as everlasting sweet peas, which will regrow each year. Climbers with woody stems are termed lianas (remarkably we know that these date back as far as 18.5 million years from fossil evidence found in Panama). Climbers don't need to grow strong trunks to support their weight, thus saving resources, and can send their stems along the ground before climbing up to the light. In the tropics, about 40 per cent of forest species are climbers.

While trees support the growth of lianas, many animals feed of lianas, eating their leaves, fruit, sap, nectar and pollen.

Tendrils are present on a range of vines, such as members of the cucumber family, passionfruit and grape vines. Some young climbers, including sweet peas, benefit from their stems being attached to supports when they are young plants; they use leaflets adapted as tendrils to reach out until they sense a support, then start coiling around it. This is called thigmotropism – the side of the stem that is being touched is slower to grow than the opposite side, thus causing the stems to curl around an object.

Charles Darwin observed how quickly cucumbers responded to gentle touch. You can test this out yourself with cucumber, pea or sweet pea tendrils! When the tendrils contact an object, they create helical coils like the wire in a telephone handset. A turn in one direction is countered by a turn in the other direction to ensure that the tendril grows in a straight line. Researchers at Harvard University found that if you stretch a mature cucumber tendril, it creates more helical loops, instead of unwinding. 'Overwinding' allows tendrils to respond not only to gentle movements but also to strong winds without breaking and the stems falling to the ground. This stiffened tendril structure has even provided inspiration for new kinds of industrial springs.

Other flowers are also known for climbing. Passionfruit (*Passiflora* species) (see page 89) can twine their inflorescences (flower clusters) around supports. Clematis stems climb using young leaf petioles rather than tendrils to twist around supports, and vigorous clematis, such as old man's beard (*Clematis vitalba*), can become an invasive liana in the wild, climbing through trees and rooting where the branches touch the ground. Some climbers use adhesive to stick to surfaces. For example, Virginia creeper (*Parthenocissus henryana*) has tendrils with adhesive pads. Some plants produce a glue from adapted roots with which to cling onto cracks and crevices. Hydrangeas are one example. Another is ivy (*Hedera*), whose thin, yellow adhesive contains tiny balls of sugar-coated protein. Once the water evaporates, calcium and pectin allow the glue to dry, thus providing very firm support. Other climbing species use hooks, prickles (like roses) or spines (like rattan palms), to scramble over other plants.

The quest for blue dahlias

Humans strive to achieve the impossible. There are no such things as blue dahlias, so in 1846 the Caledonian Horticultural Society of Edinburgh offered £2,000 to anyone who could produce one. The prize is still unclaimed.

The substances that mainly determine the colour of any flower are two groups of natural pigments. One group is the anthocyanins, which produce white, red, blue, yellow, purple and even black and brown. The other group is the carotenoids, which are responsible for some yellows, oranges and reds. There are other influences, too, including the amount of light the flowers can absorb while growing; the temperature of the air and the soil; the pH (acidity or alkalinity) of the soil; and the possible effects of stress (floods and droughts, for example).

Anthocyanins are the most important pigments for pure blue flowers. Although some plant families have evolved to produce blue flowers – cornflowers, delphiniums, gentians and pansies are among the best-known examples – many other species lack

Many attempts have been made to grow a true blue dahlia, but none has ever succeeded.

the ability to make delphinidin, a key anthocyanin for blueness. Since few plants can produce this pigment, blue flowers are relatively rare in nature: they are most commonly found in environmentally impoverished habitats, where pollinators are scarce – with its short wavelengh, the colour blue stands out and attracts bee pollinators.

Breeders of dahlias and roses have responded to the challenge of producing a perfect blue flower colour by natural hybridization. Results so far have been variable – most of the cultivars are bluish-grey and mauve-purple rather than blue – but some of the best dahlia varieties include Worton's Blue Streak, Bluetiful and Lilac Time.

There appeared to be a link in traditional rose breeding between magenta, lavender and brown flower colours. Similar hybridizing approaches to breeding blueish roses using diverse parents have been used by different breeders, but the fact remains that, in spite of their names, roses such as Blue Moon and Rhapsody in Blue are not a true blue colour. In the early 2000s in Japan, small amounts of rosacyanins – blue-mauve anthocyanin pigments – were found in the petals of the hybrid tea rose Madame Violet, and this discovery has helped to focus subsequent breeding work.

In a further effort to create a truly blue flower, plant geneticists have attempted to introduce the missing anthocyanin pigment delphinidin from blue-flowered plant species and combine it with a flavonoid enzyme. This method was first tried on tobacco and petunias, then on flower crops. Again, complete success has been difficult to achieve and results varied according to the pre-existent genes of both plants – donor and recipient – and the pH of the flower tissue.

The transformation process is complex, involving the introduction of several genes. A Moon series of carnations in a range of blue hues has been produced by Florigene, an Australian biotechnology company. The initial blue rose variety Suntory Applause incorporated blue genes from pansies, but the acidity of the flower tissue and the incomplete silencing of existing flower pigments resulted in a flower that is blue but not true blue.

Another blueish dahlia flower produced in Japan incorporates the blue anthocyanin genes of the Asiatic dayflower (*Commelina communis*). Pure blue chrysanthemum flowers have been produced in a two-step process incorporating genes from Canterbury bells (*Campanula medium*) and from the butterfly pea (*Clitoria ternatea*). This process transformed red or pink flowers to magenta and then blue. No additional genes were needed due to pre-existing colourless compounds in the chrysanthemum flower. This was the first time that such a truly blue colour flower was generated using genetic engineering technology.

Dandelion seeds have been known to travel as far as 8 km (5 miles) from the parent plant.

How the dandelion stayed alive

The sight of 'dandelion clocks' on a sunny day may tempt children to pick them carefully and then, with gentle puffs, blow the fluffy seed heads away and pretend to tell the time: thus they help to spread these plants.

Dandelions (*Taraxacum officinalis*) are members of the Asteraceae family. There are 60 macrospecies and around 2,800 microspecies – the most widely distributed are the common and red-seeded dandelions, which have spread around the globe thanks to their fluffy seed heads. They flower in profusion from early to mid-summer, then sporadically until autumn, providing bees and butterflies with important sources of nectar and pollen.

Each seed forms a 'beak' attached to a thin tube surrounded by around 100 fibres, known as pappi, which act like a parachute. They are very uniform in weight and geometry wherever they are produced. The dandelion's seed clock can control seed dispersal to suit its needs, so the hairs open only when the weather is dry and more likely to be breezy.

Wind-tunnel experiments show that the dandelion's pappi create a circle of air – a 'separation vortex ring' – above the fluffy hairs, slowing the seeds' descent and thereby increasing the likelihood of a gust of wind carrying them up and away. The pappi are 90 per cent porous, letting air through the hairs and causing four times less drag than a solid disc of the same size. It seems that this advantage works only because the dandelion seed is very light. This means that the seeds can travel up to 100 km (60 miles), though some earlier research indicated that 99 per cent of seeds fall within 10 m (30 ft) of the parent plant. Each dandelion plant can produce about 2,000 seeds from around ten flower heads.

Dandelions are incredibly successful at colonizing land. Seeds can also be dispersed by water. It is estimated that more than 97 million seeds per hectare (2.5 acres) could be produced yearly by a dense stand of dandelions. The largest proportion of seeds are produced and germinate in the same year. Most are short-lived, but some can last for about five years if buried in soil.

Another epic flier in terms of seed dispersal is the Javan cucumber (*Alsomitra macrocarpa*). The vines of this member of the Cucurbitaceae family scramble up into the tropical tree canopy and produce bell-shaped fruit 30 cm (1 ft) wide, inside which winged seeds develop. As the fruit dries out at maturity, the seeds drop out of a hole in the bottom of them. Many fall straight to the ground, but the seeds' transparent, thin and flexible swept-back wings of around 13 cm (5 in) in diameter facilitate amazing pitch and yaw control in the slightest air currents and may result in dispersal over much greater distances – they have even been found on ships out at sea!

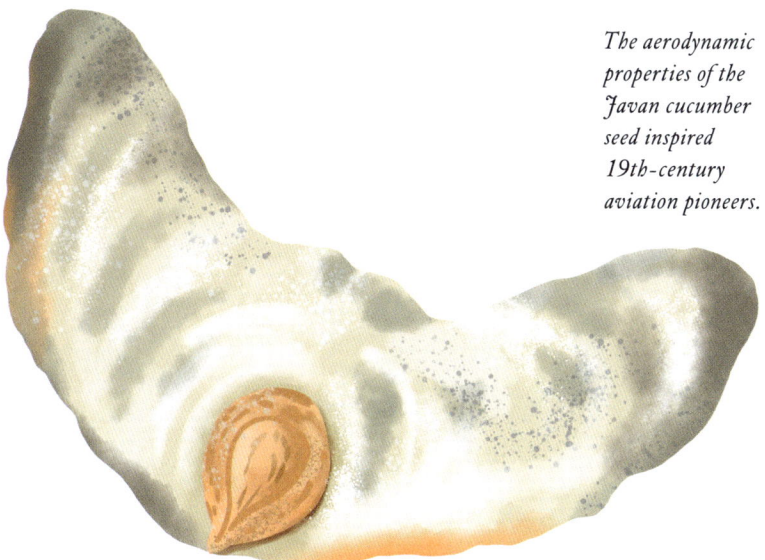

The aerodynamic properties of the Javan cucumber seed inspired 19th-century aviation pioneers.

In 2022 a 'transient robotic' biodegradable sensor was developed based on the aerodynamics of these flying seeds for mass release from aircraft for ecological surveys. The sensor's flight mimics the seeds' spiral motion, landing the glider close to the launch position. The wings biodegrade in seven days, and a biodegradable cellulose gelatine layer covering the rainwater pH-sensing layer provides visual feedback. The flight mimicked that of the Javan cucumber seed – spiral motion acts as aerial braking, landing the glider close to the launch position.

The orchids that grew on the tree

Orchids are epiphytes (plants that grow on other plants, but not as parasites). Some are easy to care for, and hence make popular house and garden plants. Others are complex and difficult to grow in cultivation.

Moth orchids (*Phalaenopsis*) and corsage orchids (*Cattleya*) are very adaptable to artificial light in our homes as long as they have the right watering and feeding regimes. Of course, that is true to some extent of most flowers, but orchids are a little more demanding – not a lot, though: they are happy to grow in pots of special free-draining orchid compost or attach their root systems to strategically positioned wooden slabs.

In tropical or subtropical forests, the area between the ground and the canopy has niches for plants adapted to dappled lighting or shade to grow in, and it is there that epiphytes most prefer. Monocot species (see page 20), including orchids, are the most common group of epiphytes. Cracks, fissures and tree branches provide ideal habitats for epiphytic orchids, but, unlike

Moth orchids, which do not need soil to survive, are native to Southeast Asia and Australia.

parasites, they don't penetrate the bark – they capture nutrition from the air, dust, moisture, pockets of humus, decaying plant material or dead animals caught on the bark of the host tree or shrub. Although this falling debris helps them grow and thrive, it doesn't always benefit their appearance: these orchids are generally less showy and almost invariably more ragged than specimens kept indoors or in botanical gardens.

Some wild orchids have adapted to grow in ant colonies, where they benefit from nutrients gathered by the insects, and sometimes have sticky seed coats that attract the ants to disperse their seeds. Many orchid species must adapt to wide ranges of daily temperature, and in some cases to harsh, seasonally arid conditions. To cope with dry conditions, some orchids hold water in rhizomes (rootstalks) to ensure their survival; it's a hard life because, unlike terrestrial plants, they don't have deep layers of soil to rely on.

Orchids face a particularly daunting survival challenge because the tiny seeds they produce contain little or no nutrition. Germinating orchid seeds do not photosynthesize initially; instead, they form relationships with mycorrhizal fungi, which are vital for their establishment and development. The fungi provide the young orchids with a source of carbon, nitrogen and phosphorus from decaying wood, or tapped from the living host tissue. The fungi form hyphal loops or 'pelotons' within the orchid root.

Some relationships between orchids and fungi are mutually beneficial; others are good for only one of the parties. Some orchids form long-term relationships with one particular fungal species; others switch between one or more species as they grow from seedlings to adult plants. This change of attachment may be essential when the original wood-decaying fungi have finished their work. Alternatively, in environmentally impoverished growing conditions, orchid species cannot be so fussy: they have to rely on a wide range of fungal species in order to survive.

The pace of life is slower in temperate forests than in the tropics, and there are fewer orchids – indeed, fewer epiphytes of any kind – here. In old forests, there is a need to replant new trees to provide a continuity of habitat that benefits not only the trees but also the epiphytes that grow on them. In Denmark, old colonies of European lady's slipper orchids (*Cypripedium calceolus*) growing in ancient beech (*Fagus sylvatica*) forests are ageing and flowering less, and their numbers are declining because there are no new trees for their seeds to colonize. Habitat conservation and restoration are extremely important to support orchids growing on trees in different parts of the world.

The flower structure of the arum lily is similar to that of the titan arum, but fortunately it lacks the latter's stink.

How the arum flower grew so big

Arums are renowned for their large, trumpet-shaped flowers. They are the confidence tricksters of the plant world because they attract insects searching for food but then don't give them anything to eat.

In the rainforests of Borneo reside two of the world's strangest plants producing giant flowering structures. Both are the colour of rotting meat, and both give off the smell of putrefaction, but they are unrelated. One produces the biggest inflorescence (flower head) in the world, at around 2.7 m (9 ft) in height and 1 m (3 ft) across. It consists of a giant spathe or sheath protecting a flowering stem called the spadix that supports many small separate male and female flowers. This is the titan arum (*Amorphophallus titanum*). The arums most frequently seen in gardens worldwide are cultivars of the arum lily (*Zantedeschia aethiopica*).

The other oddity is the parasite *Rafflesia arnoldii*, which lives on the Indian chestnut (*Tetrastigma leucostaphylum*), a climbing vine.

This has the largest individual flower in the world, at more than 1 m (3 ft) across. It develops from a flower bud emerging from the ground and opens to produce either male or female flowers after nine months of development. Until that bloom appears, there is no sign of its presence: it's a complete parasite that doesn't photosynthesize.

The titan arum grows in rainforest clearances. It is a perennial that produces large rhizomes (rootstalks) to store energy. About ten years after germination enough energy is produced for it to flower. The leaf withers and the giant flower bud emerges from the soil. At dusk, when carrion insects are active, it heats up and opens its flower to emit a pungent smell similar to that of rotting meat. The aroma spreads over a wide area and attracts sweat bees (bees attracted to salt in human perspiration) and other carrion insects. Some of these visitors – which may be covered in pollen from other titan arums – enter the spadix looking for a location to lay their eggs. Instead, they find themselves trapped in a lower chamber by hairs below the male flowers and by constriction of the spathe. They remain in captivity throughout the night, during which pollination takes place. When that's complete, the hairs wither and the insects can crawl out past the now ripe male flowers, picking up pollen en route to pollinate other titan arums.

Rafflesia plants also generate heat along with a noxious smell when their flowers open. The flowers actually resemble rotting meat in colour, texture and patterning to doubly attract carrion insects as pollinators. Although some rafflesia species have flowers with both male and female parts, this isn't the case with *R. arnoldii*. This means that carrion insects lay their eggs inside male flowers, becoming covered in pollen in the process, and the plant has to rely on them flying off to find female flowers so that pollination takes place. The pollen lasts for several days – ample time for the insects to find female flowers – which is just as well, since *R. arnoldii* is a rare plant threatened by deforestation.

In North America, the skunk cabbage (*Symplocarpus foetidus*) is an early spring arum. The flower's carrion-like odour – reminiscent of the mammal from which it gets its name – together with the heat generated by the flower when it opens – enough to melt surrounding snow – attracts blowflies, which together with other early-spring insects (gnats, bees and beetles) act as pollinators. It also drops pea-sized seeds from its decaying fruit head which are spread by birds and small animals that eat then defecate them.

How the aubretia learned to crawl

One of the odd things about aubretia is its name, which ought to be 'aubrieta', after French painter Claude Aubriet, but outside botanical circles it is commonly spelt and pronounced 'aubretia'.

The *Aubrieta* genus contains around 20 species of flowering plants in the cabbage family (Brassicaceae). It is known as a decumbent or a prostrate – in other words, its branches spread horizontally before turning towards the light. This is unusual in nature but by no means unique: among the other plants with the same growth pattern is the trailing wakerobin (*Trillium decumbens*), a herbaceous perennial of the southeastern USA, whose stems crawl along the ground before growing upright.

Other decumbents include ivy (*Hedera helix*), which can spread along the ground and climb up fences and trees, and the lesser periwinkle (*Vinca minor*). If unchecked, such species can run wild, but they also have some uses: creeping juniper (*Juniperus horizontalis*) may be cultivated to bond soil and prevent erosion.

Aubretia grow sideways first, then upwards – an ideal approach to challenging conditions.

Some prostrate plants are conspicuously successful at preventing competition from other species: a leading example is creeping Jenny (*Lysimachia nummularia*), which, as its name suggests, is a creeper.

Generally, decumbents and prostrates are found in poor soils or in shade. Their leaves sometimes provide colourful ground cover: the small, hooded magenta or white flowers of the spotted dead nettle (*Lamium maculatum*) may carpet moist, shaded areas of woodland; also attractive is the similarly coloured creeping phlox (*Phlox stolonifera*). If gardeners like such species, all well and good; if they do not, they are easy enough to restrain by pulling back exuberant growths and snipping them off.

Aubretia species are well adapted to dry Mediterranean, Alpine, desert or Arctic conditions – their low growing habits are an advantage in tough environments. Some aubretias (in the relative luxury of an ornamental rock garden) can spread not only to form carpets, but also by seed into cracks in walls, gaps in pavements and well-drained scree slopes. Some species, such as rock cress (*Aubrieta deltoidea*), which occurs naturally in southeastern Europe and can be found in the Tartej uplands of Lebanon, have produced many garden varieties. They form low, evergreen carpets that spread out covered in typically lilac-coloured flowers.

This should not be taken to imply that all prostrates are picturesque: their method of spreading can lead to the foliage looking untidy and gappy after flowering. Again, however, this is a minor problem that can be solved with a haircut.

Although life is harsh in mountainous and Arctic regions, prostrate plants can thrive in such surroundings because their leaves help to protect them from the wind and provide an insulation blanket for the surface of the ground. In some cases, hairs on their leaves also protect them from damage. They can be adapted to tolerate a very wide range of diurnal temperatures, from very cold soil to the potentially damaging intense heat and UV light from sunlight captured by their leaves. In some cases, prostrate Alpine plants produce large flowers that can capture the heat of the sun, which helps to attract pollinators and delights alpine gardeners. These species are adapted to snow cover, which acts as an insulating blanket.

Some recumbents regrow rapidly when spring returns thanks to their evergreen leaves. This gives them a head start in short growing seasons because having foliage already present reduces the need for energy renewal. Prostrate alpines growing under harsh conditions are very conservative – they don't put on lots of growth each year. Moreover, they recycle their nutrients in generally impoverished soils by utilizing the leaf litter under their prostrate canopy for regrowth – the low canopy helps to prevent debris from being washed away down the mountainside.

Red poppies are annuals, but other members of the Papaver genus are binennials and perennials.

The red poppy that lasts only a year

The red poppy (*Papaver rhoeas*) is a much-loved wildflower. It's an annual, which means that it completes its life cycle – from germination to the production of seeds – within a single growing season, and then dies. Sadly, it sheds its petals quickly when picked.

Red poppies are ideal for introducing young gardeners to the delights of growing flowers from seed, either as a mass of red flowers on their own, as varieties selected for mixed flower colours, such as Shirley Poppies – selected originally by a vicar in Shirley, a suburb of Birmingham, UK – or in wildflower mixtures. They grow in cultivated or disturbed ground and flower rapidly the same season, producing many seed pods and seeds. When the pods ripen, they rattle when shaken and the seeds are distributed as the pods are blown in the wind.

The common names for the plant, such as corn poppy and field poppy, tell you a lot about their natural habitat. They grow typically on disturbed or cultivated land, completing their life

The leaves of the flower-of-an-hour may have an iridescence that, though invisible to humans, is attractive to bees.

cycle along with crops or wayside weeds, and then distributing their seeds to build up banks of seed in the soil to be grown in subsequent years. They originated in southern Europe and have since spread around the temperate world. They were once much more common, adding splashes of red to cereal fields. Until the advent of more intense farming practices and chemical herbicides, poppies were a rich source of pollen for insects.

Corn poppies have a lasting symbolic link with the fields of Flanders during the First World War – where the land churned up by the conflict resulted in the appearance of a sea of scarlet flowers from the long-buried banks of poppy seeds brought to the surface by the attrition of war. Poppies feature in one of the best-known First World War poems, 'In Flanders Field' by John McCrae, and the flowers are now immortalized as emblems of remembrance for those killed in armed conflicts everywhere.

Once you have poppies, they will spring up from seed every year in the most unlikely places unless weeded out on a regular basis. However, modern intensive production systems with herbicides and fertilizers have made their traditional crop habitats rare. Today poppies are more likely to be seen on organic cereal farms but, along with another weed, charlock (*Sinapis arvensis*), they impact on farmers' cereal yield, so they are often weeded out by a specially designed robotic tine. Meanwhile, there are conservation projects elsewhere to maintain traditional cropping systems for field weed species to ensure that biodiverse habitats that include poppies survive for the benefit of wildlife.

While poppies are quick to grow, they are outpaced in all phases of their life cycle by *Hibiscus trionum*, whose common name is highly telling: it is widely known as flower-of-an-hour, which is an exaggeration but not a big one. It is a garden-worthy annual with attractive flowers but which unfortunately also produces prolific amounts of seeds and spreads to disturbed wasteland, cultivated farmland and garden plots.

The apple tree that didn't fruit

Fruit trees cannot always be relied on to produce a crop annually. If missing a year meant that they were no longer fertile, that would be easy to deal with, but trees may miss one year, then fruit prolifically the next.

There is a wide range of perennial and annual plants that vary in yield from year to year. This is because the amount of energy required to produce a heavy fruit crop taxes the tree, which must then rest and recuperate by producing fewer flowers and fruit. One year 'on' and the following year 'off' is termed biennial bearing, but to complicate matters, some trees may miss more than one year before recovering their fertility. Apples, Arabica coffee, apricots, avocados, pears and mangos are biennial bearers; grapes in general are not.

The incidence of biennial bearing can depend on the variety of the particular fruit tree. Genetic make-up affects flower onset and crop load; heat accumulation may affect flowering the following year. Plant hormones – particularly levels of

The appearance of apple blossom is not a reliable sign that fruit is on the way: some trees flower, then bear no apples until the following year.

gibberellin in developing seed embryos – can suppress flower bud development, and scientists have been assessing the number of seeds fruit produce to see if this has any effect. Sugar and carbohydrate levels can also have an impact: understanding the genetic controls of this may lead to breeding for guaranteed yearly fruit production. A few years ago, there was interest in Ballerina apple trees – derived from a mutation in a McIntosh apple variety that had short fruiting branches and didn't need pruning, ideal for commercial orchards. Unfortunately, this was linked to biennial fruit bearing. As some apple tree varieties age, they become more biennial in bearing fruit. This unpredictability has commercial implications for farmers, not only in terms of year-to-year yield, but also in terms of size of fruit, since heavy crop loads in 'on' years can result in smaller fruit, unless they are thinned by hand, which is an additional expense. Pruning, correct irrigation, use of fertilizers and selection of varieties that fruit every year can overcome some, if not all, of these problems.

Apples varieties may vary: Royal Galas regularly bear fruit each year, but Fuji dessert apples and Bramley cooking apples tend to bear fruit biennially. This may be genetic, but climate change is also a factor: some types of apple are no longer getting the degree of winter chill to which they were accustomed. As a result, in addition to their usual spring flowering, they are trying to flower again just before leaf fall; this has a knock-on effect to the following year's crop, which tends to be less colourful and smaller in both size and number. In response to global warming, scientists are working on blackcurrant breeding programmes to reduce that fruit's need for cold weather to produce flower buds.

Coffee producers also have a problem with biennial bearing, and this is independent of weather and climate. In Ethiopia and other parts of East Africa, output in an 'off' year can be between five and ten times less than an 'on' year. In 'on' years, the plant's resources are channelled into flowering and fruiting at the expense of vegetative growth; in 'off' years, the latter takes precedence over the former. One theory is that in nature this is a way of controlling predator populations that impact seed production and species survival.

Nut- and seed-bearing trees also have synchronized cycles of good and bad years. Both beech and oak alternate between bumper seedings – known as 'mast years' – and poor returns in the years in between. This appears to be affected by pollen production levels, the populations of predators that specialize in eating the seeds, and the level of survival of seedling trees from previous mast years. If lots of seedlings survive in the understorey of the woods, tree species are more likely to have 'on' and 'off' years because there is less pressure to produce replacement seedling trees annually. The seedlings remain small until a gap occurs, allowing them to grow further and fill the available space.

Because petunias have high growth points, they are easy to uproot unintentionally.

How the grass was mown, but not the petunia

If you mow the lawn, the grass can usually be relied on to grow back; but a petunia can't withstand regular cutting. The difference is the relative positions of their growth points: in grass it's at the base, in petunias it's higher.

This is not to suggest that lawns are easy to maintain: if they don't get enough feed, or if they're cut too short or too infrequently, they may lose their lush appearance and develop gaps in which weeds can germinate.

Some invaders, such as the daisy-like camomile (of the Asteraceae family), may eventually die out after regular cutting. Other weeds are harder to get rid of, either because they are too low to be cut or because they have food stored deep in their roots. Among the most unwelcome species are the broad-leaved creeper speedwell (*Veronica*), deep-rooted dandelions (*Taraxacum*) and docks (*Rumex obtusifolius*), all of which will need at least a daisy grubber to remove, and possibly a spot herbicide treatment. (Prevention is better than cure: it's best if possible to plant hard-wearing grass and clover seed varieties, to cut the lawn lightly and often – not too short – and to feed regularly to create a thicker grass canopy.)

For golf courses and permanent pastures, it's best to install grasses that are low-growing or have a horizontal growing habit. Short species, such as red fescue (*Festuca rubra*) are used for golfing greens to withstand close cutting; however, these are not very hard-wearing unless grown in mixtures with other, more robust species. Sheep's fescue (*Festuca ovina*) requires minimum year-round mowing: it isn't hard-wearing, but it may be suitable for grass banks that don't need cutting at all.

Mixtures that include perennial rye grass (*Lolium perenne*) provide tougher-wearing lawns suitable for the kind of back gardens that are used for impromptu games of football. These plants tend to have a prostrate habit, so spread out sideways, with their growing points close to the ground, safely beneath the blades of the mower.

Many of the hard-wearing requirements for sports turf and back gardens are the same as those for long-term grazing meadows, where the need is for grass that forms wide clumps, covers all the ground and has sufficient bulk for vertical growth for grazing in spring and summer.

Farmers sometimes sow short-term lays. These are fields put out to grass for a few years before they are ploughed and planted with crops. For this purpose, Italian rye grass (*Lolium multiflorum*) may be preferred to perennial rye grass: the Italian variety is more upright-growing, it doesn't form a prostrate mat, it starts growing earlier and continues later in the year. Moreover, it produces a greater yield per season and can be cut several times a year. But since the growing points are higher off the ground, Italian rye grass doesn't survive as long as prostrate perennial rye grass.

Similar variations in prostrate to upright plant habit exist in the major clover species, such as white, red and crimson clover, used for lawns, grazing and conservation or cover crops to protect the soil. The clover complements the grass, and its roots fix nitrogen from bacteria in root nodules.

In Britain since the 1930s, some 97 per cent of wildflower grassland has been lost. There is now renewed interest in cutting grassland to encourage flowering plants. If biodiversity is low, two cuts a year will be required after seed set, and strips of uncut foliage should be left as refuges for insects. It is important to remove the cut foliage to lower the fertility of the site and thus encourage flowering plants in preference to grass and clover.

How the sunflower tracked the sun

If you're on the beach trying to get a tan, you might move around as the sun's position in the sky changes from east to west. Some plants do something similar: although they can't walk, they follow the sun with their heads.

In ancient times and throughout the Middle Ages it was widely believed that heliotropism – a tendency to move in the direction of the sun – in plants was entirely passive: in other words, that the movement was none of the flowers' doing; they were merely drawn towards Earth's star in much the same way as ocean tides are influenced by the moon.

However, research in the 19th century demonstrated that, on the contrary, this phenomenon is plant-led: having followed the course of the sun during the day, sunflowers (*Helianthus annuus*) and some other plants reposition themselves during the night so that they face east in readiness to catch the earliest rays at dawn the following day.

Sunflowers track the direction of the sun in order to warm their flower heads and attract pollinators.

Sunflowers have pulvini (singular: pulvinus) – swellings at the base of their leaves that resemble inflatable cushions. These act like joints, enabling the plants to move in directions other than those determined by normal growth.

Closely related to heliotropism is paraheliotropism – a plant's ability to respond to excessive heat by reorienting itself to reduce the surface area of its leaves that are in direct sunlight. Among the plants that can do this are cotton (*Gossypium*) and soybean (*Glycine max*).

Further study of sunflowers has shown that their stems elongate unevenly. During daylight hours, there are growth spurts on the side of the stem that faces the sun, and very little elongation on the other side. At night, the opposite occurs: the side of the stem that was further from the light source during the day grows faster than the other side, thus causing the flower head to return to face east in anticipation of the dawn. When the sun rises, the flower heads warm up very quickly, thus attracting around five times more pollinators than non-heliotropic plants – which warm up later in the day – and encouraging the visitors to stay longer than they might if there were more competing plant species nearby ready to receive them.

Flowers of the common daisy (*Bellis perennis*) also track the sun. During daylight hours, their movements, like those of the sunflower, are controlled by pulvini, but at night, when the daisies' heads close up, their orientation is random. They open up again when the sun rises the following morning, then keep their heads pointing at it until it sinks back below the horizon.

It is from this behaviour – opening at dawn and closing at dusk – that they get their common English name, which is a conflation of 'day's eye'.

In a variation on the heliotropic theme, agapanthus flower buds bend towards the sun until the early afternoon, before moving back to their original position by late afternoon. Some Arctic and Alpine species also track the sun: flowers of the Arctic poppy (*Papaver radicatum*) face the sun for 24 hours a day in midsummer. The South African Cape daisy (*Osteospermum*) opens its blooms only in sunshine or other bright light, keeping them closed at night and on overcast days. Before it comes into leaf, the flowers of the North American white trillium (*Trillium grandiflorum*) face mainly south, in order to benefit from the warmth of the sun and high levels of insect fertilization, but daily variations in the horizontal and vertical orientation of some of its flowers have also been observed. Some forms of morning glory – the common name for more than 1,000 species of flowering plants in the family Convolvulaceae – follow the sun in a general way but not exactly, often because day-long focused concentration would be damaging in hot climates.

The daisy is one of the most adaptable flowering plants in the world.

How the daisy became so common

Daisies are members of the *Asteraceae* family, perhaps the world's most successful family of flowering plants. They can adapt to almost any environment, and flourish in earth that has been disturbed by human activity.

Almost 95 per cent of all plants are flowering plants, and of those about one tenth are Asteraceae – between 25,000 and 35,000 species within 1,700 genera. Asteraceae are found in a wide range of climatic conditions and habitats, from the tropics to the temperate zones.

There is a wide range of highly ornamental species and varieties to suit gardens in many parts of the world: among the most famous are asters, chrysanthemums and dahlias. Asteraceae provide humanity with food – artichokes, lettuce and sunflowers are family members. They are grown as herbs (tarragon) and used as medicine (camomile). Extracts from the dried flowers of another genus, *Chrysanthemum* or *Tanacetum*, contain a chemical that stops mosquitoes from biting and controls crop pests.

Asteraceae also provide food for birds feeding on their seeds, and some types attract insect pollination. Goldenrod (*Solidago*) and knapweeds (*Centaurea*) are good sources of nectar and pollen for honey-bees.

Much of the success of the *Asteraceae* is down to their reproductive and seed dispersal system. Their flower heads are composites made up of a series of individual flowers or florets within the flower head (florets around or within the flower head often have petals, which are the attractively coloured part of the flower).

When ripe, the seeds of the Asteraceae family are light and dry, which makes them easy to disperse on the wind, and surrounded by tufts of hair, which make them unpalatable to eat. Some species, such burdock (*Arctium*), have evolved burrs (hooks) around their seeds, or complete florets that get distributed by attaching themselves to animals brushing past them. The seeds themselves are achenes – dry fruits that are inhidescent (do not split open when ripe) and in which the single seed is separate from the fruit wall. Achenes have adapted to a range of different dispersal mechanisms: some of them float downstream until they become waterlogged; others are distributed on birds' feet or by ants, which in certain cases are attracted by a free meal from part of the achene wall.

The origins and global spread of the daisy family are gradually becoming better understood thanks to recently discovered archaeological remains and modern genome analysis techniques. It appears that the daisy family arose when early flowering plants were developing 83 million years ago in South America (remarkably, the oldest lineage, the subfamily Barnadesieae, which contains 90 species, still exists). At this stage, dinosaurs were still roaming the Earth, birds were increasing in numbers and mammals were generally still quite small. While America was moving away from Europe, Australia was joined to Antarctica and moving away from Latin America and Africa. There was a mass extinction event about 66 million years ago. Currently it is thought that the Asteraceae moved via North America and Asia to Africa, arriving there about 42 million years ago in the Eocene (New Dawn).

This was a period of dramatic change from predominantly tropical forests and high carbon dioxide levels worldwide to lower temperatures and reduced amounts of carbon dioxide and a period of extensive diversification in the Asteraceae resulting in many of the daisy species we know today. During the late Oligocene, roughly 25 million years ago, there was further diversification in the sunflower tribe while the world experienced even cooler temperatures. Asteraceae have spread around the world from Africa to be the successful family that they are today. By contrast, modern humans were not on the scene until about 300,000 years ago!

How the oak was galled

Most times when you walk beneath an oak tree, the only sound you'll hear is the rustling of leaves. The tranquillity hides the fight for survival among the 2,300 species of organisms that may be living within and on it.

Oak trees in the UK can support 38 bird species, 229 bryophytes (mosses), 108 fungi, 1,178 invertebrates, 716 lichens and 31 mammals – not to mention numerous bacteria and microorganisms. Among the insects reliant on oak trees are up to 70 species of oak gall wasp. Worldwide there are about 1,000 species of oak gall wasp, with the number of species in a region depending on the diversity of oak trees. For example, in North America as far south as northern Mexico there are around 700 species. Oak species with higher tannin levels support a greater diversity of gall wasps (the tannins have some protective effect against fungal infection of developing grubs).

Although the word 'gall' sometimes has undesirable connations – it's bile in a human, a painful swelling in a horse – none of

A gall on the branch of an oak tree may
be mistaken for a parasite or a disease,
but it is neither.

them applies to these insects, which are part of the oak tree's biodiversity: they have little or no impact on its health and growth, so it is undesirable to attempt to control them.

Each oak gall wasp flies to its target site on the oak tree – including leaves, buds, stems, catkins or female flowers that will develop into acorns – and lays an egg. Species that lay eggs in developing flowers can markedly affect both the quality and quantity of the acorn crop. The gall wasp's eggs and grubs produce secretions that induce a chemical reaction in the oak tree to develop a growth that protects the young grub – in effect, the host tree builds a shell around the insect. This develops into a chamber into which the grub moves after hatching. As the grub feeds on the plant tissue inside the chamber, its secretions continue to direct the oak's tissue to develop the small chamber into a larger gall. The plant produces thin-walled nutritive cells, such as seed tissue, inside the gall to feed the growing grub. Some galls fall to the ground in the autumn, and it is important that they fall before the leaves so that they can be camouflaged by them on the ground.

The developing galls are specific to the particular gall wasp and come in lots of different shapes – they may resemble oak apples (which may be brownish, yellowish, greenish, pinkish, or reddish), oak marbles, oak artichokes or acorn cup galls. Knopper galls are distortions of developing acorns on pedunculate oak trees (*Quercus robur*). (Peduncles are stalks.) Some galls have more than one chamber, and not all the grubs that develop therein are from the same female – some gall wasps are opportunists, laying their eggs near other insects' developing chambers.

A parasitic gall wasp laying an egg in a gall.

The developing gall should protect the grub, but it's not all plain sailing: there are lots of threats still to be confronted. Parasitoid gall wasps can drill through galls with their ovipositors to lay eggs in the developing gall wasp eggs; moth larvae can predate developing gall wasp larvae in their chambers; fungi can attack certain species of gall wasp larvae.

Galls have several distinct forms, all of which are designed to ward off predators, especially to reduce the possibility of parasitic wasps laying their eggs in grubs inside the chamber. Among the protective devices are thick, tough gall cases, spines, sticky resin coatings and nectar to attract defensive ants.

Houseleeks – sometimes known as liveforevers – are ideal for planting on roofs.

How the houseleek got on the roof

Succulent plants are adapted to some of the world's harshest environments. They can grow in poor, thin, well-drained soil, on rocky mountain slopes in cracks and fissures, and in both alpine and sub-alpine zones.

There is no one family of succulents: they occur in nearly 60 families worldwide, divided into more than 300 genera, including *Crassula*, *Haworthia* and *Sedum*. They are superficially similar to cacti, but they lack the areoles (small lumps or spines) that characterize the latter. Cacti store water in their fleshy stems, whereas succulents hold water and nutrients in their leaves. The North American succulents that most closely resemble cacti have thorns rather than areoles, and produce toxic white latex from their stems, which no cactus can do.

Again unlike most cacti, which require dry environments, some succulents can cope with winter rain and cold, and even snowfall, as long as they do not become waterlogged. Their sometimes effortless-seeming ability to survive uprooting and

177

Echinocactus *is
one of the most
popular cacti
for domestic
cultivation.*

replanting has inspired their Latin name *Sempervivum*, which
means 'always alive'. One species in particular, the houseleek
(*Sempervivum tectorum*), is abundant throughout southern
and eastern Europe, North Africa and western Asia, where it
is often found in small moist cracks on the sides and tops of
buildings (*tectorum* is Latin, meaning 'of the roofs'). In such
locations it has long been actively encouraged: the Holy Roman
Emperor Charlemagne (*c.*747–814) decreed that houseleeks
should be planted on roofs to strengthen them and protect them
from decay. That practice was revived in eco-conscious 1960s'
Germany, where houseleeks were planted alone or in mixtures
with other succulents and alpine species on office blocks and
houses. In the new millennium green roofs are appearing in
cities all over the world, as their benefits are now well known:
they cleanse rainfall and slow its run-off; they provide a habitat
for birds and insects; they help to improve the quality of urban
air; they reduce noise pollution; and they provide potential
firebreaks between adjacent buildings.

One of the most adaptable and attractive ornamental succulents used for green roofs is the myrtle spurge (*Euphorbia myrsinites*), which has trailing stems that can be cut back after flowering. The species has spiralling, geometrically arranged blue-grey leaves, and yellow flower clusters from May to August. It is native to southern and southeastern Europe and Asia Minor (now called Anatolia), adapts well to many conditions – sunshine or partial shade; moist to dry well-drained soils – and is grown with other succulents and alpines in gardens from California to Europe.

The parent rosettes of succulents produce small, satellite rosettes that radiate out to form new offset plants. After about two years, the main rosette runs up to flower and then dies, producing fine seed. A diverse range of flowers and coloured and 'cobweb' leaf forms have been produced in cultivation to brighten up green roofs throughout the year.

In spite of its name, the houseleek is unrelated to vegetable leeks, which are members of the *Allium* genus, along with garlic, onion and shallot.

The greater celandine that talked to the lesser celandine

Some plants can have almost identical common names but be very different species. In some cases, plants can look similar but not be related, or they may have several different common names from place to place in a single language.

At first glance, the greater celandine (*Chelidonium majus*) looks like the lesser celandine (*Ficaria verna*). The lesser celandine has bright yellow flowers and is a magnet for early pollinating insects. It is a harbinger of spring, with a reputation for flowering around 21 February – named 'Celandine Day' since the Reverend Gilbert White in Selbourne, Hampshire, UK, recorded it flowering on that day in 1795. By contrast, greater celandine is a taller woodland, roadside and waste-ground plant that flowers between April and October. Both plants have common names derived from *chelidon*, the Greek for 'swallow' (the bird). Superficial visual similarities are misleading as the two plants are unrelated: greater celandine is a member of the poppy family (Papaveraceae) while lesser celandine is a member of the buttercup family (Ranunculacae).

Greater celandine has a long history of use as a herbal treatment for a wide range of maladies, from indigestion to toothache.

Other plant names can be similarly misleading or confusing, especially when (as often happens) a single plant is known as more than one thing in the same language. Take, for example, the leopard lily (see pages 24–27): is it a tropical house plant with blotchy leaves, an iris, a lily, or a spotted leaf African hyacinth? The answer is that it may be any of these things – the tropical house plant is also known as dumb cane; the iris could be a blackberry; the lily is the panther lily; and the African hyacinth is also know as silver squill.

So in order to determine accurately which plant is which, we need better and further particulars than the common name. This is possible thanks to the work of Swedish botanist Carl Linnaeus (1707–78), whose book *Systema Naturae* grouped plants and animals into five categories in a system that is still used today. The most general category is 'Kingdom': this group has five subheadings: Animals; Plants; Fungi; Protista (organisms with more than one cell that don't fit into any of the previous subheadings); and Monera (single-cell organisms). Plants are then further subdivided into four more categories; these are, in increasing order of specificity: Class, Family, Genus and Species.

To illustrate the system, take the lesser celandine. First and above all, it's a plant (Kingdom: Plantae). It's a flowering plant (Class: Angiosperm). It's in a group of flowering plants that also contains poppies and buttercups (Family: Ranunculaceae). It's more like one particular group of buttercups, the *Ficaria*, so that's its Genus. Finally, lesser celandine flowers strikingly in the early months of the northern hemisphere year. Linnaeus required that all species should have two names (binomials), so lesser celandine is known botanically as *Ficaria verna* (*verna* is Latin for 'spring').

Greater celandine is also a plant (of course), and also an *Angiosperm*. It's a poppy, so it's a member of the Papaveraceae family. Its closest relatives are known as *Chelidomia*, so its Genus is *Chelidomium*. And since it's the greater, its binomial is *Chelidonium majus* (*majus* is the Latin for 'larger').

Sometimes there may be an additional name after the binomial. This identifies the person who first described the plant. Hybrid plants are classified in this way: *Clematis* x *Jackmanii*, for example, is a cross ('x') between *Clematis lanuginosa* and *Clematis viticella* bred in the mid-19th century at Jackman's Nursery in Surrey, England.

Moreover, due to natural species diversity, genetic changes or mutation, new forms can develop such as *Clematis* x *Jackmanii* '*Superba*', which has more profuse blooms than the original form. Binomial naming of plants is an internationally recognized system, but as we will discover, there are still a few challenges (see pages 204–7).

The rustling of a willow in the wind is one of the most evocative sounds of a British summer.

The willow that wept

Plants and trees don't just bend in the wind: they respond to this and other natural elements in a variety of ways; sometimes they resist, at others they adapt themselves to accommodate forces that they cannot combat.

The process by which plants and trees respond to externally applied physical or mechanical sensations is known as thigmomorphogenesis, an English word derived from the Greek for 'touch', 'shape' and 'creation'. That flora react in this way has been recognized by scientists for centuries: at the start of the 1800s, British botanist Thomas Knight found that free-standing apple trees in windy locations developed shorter, stockier stems and thicker trunk bases than examples of the same species grown in sheltered locations. Later scientific studies showed that Sitka spruce trees (*Picea sitchensis*) responded to gales in two distinct phases: first by almost immediately developing stronger roots on the side facing away from the prevailing wind to stop themselves from being flattened; and then by the following year growing comparably strong roots on the opposite side to restore their equilibrium and balance. Significantly, there was no discernible strengthening of the roots around any other parts of these trees – the growths were all along the line of the prevailing wind.

Further research confirmed earlier observations: scientists on the Pacific island of Guam who bent back the young stems of the rare Pacific Ocean tree *Serianthes nelsonii* at 90 degrees twice a day for 14 weeks found that its trunk grew stronger and shorter than other members of the same species that had not been thus manipulated: this finding is important when establishing trees in conservation projects.

As well as wind, trees may respond to a range of other stresses – such as those caused by gravity, growth surges, light variations and ripening fruit – by producing what is known as reaction wood. Reaction wood takes two main forms. In addition to thickening trees' structural elements, the form produced by broad-leaved trees increases their resistance to wind by increasing the flexibility of their branches and trunks. Known as tension wood, these growths have greater concentrations of cellulose and gelatinous fibres than normal soft wood – sometimes as much as 60 per cent higher. Tension wood develops on the upper side of a leaning tree or branch to stop it being pushed back further.

Compression wood – reaction wood produced by conifers – has less cellulose content than normal wood of the same species, and is between 15 and 40 per cent heavier because of its greater concentration of lignin, a polymer that is similar to cellulose but tougher. Compression wood develops on the lower side of a leaning branch or tree.

Reaction wood can be a problem if a straight grain is required. For example, cricket bat willows (*Salix alba caerulea*) are grown for their straight, fine-grained trunks; the best such trees are found in England and Kashmir. Unfortunately, strong winds may bend and twist the trees and cause defects in the timber, stretching cells on the outside of the bend and compressing those on the inside, thus weakening or breaking the cell structure of the wood. Consequently, cricket bats made from reaction wood are sold by some commercial makers at big discounts because of concerns about the products' strength and durability.

On the benefit side, reaction wood in willows can benefit biofuel renewable energy production: experiments conducted in the early 2000s on the Orkney Isles off the north coast of Scotland revealed that willows exposed to high winds (and therefore with a high concentration of reaction wood) produced five times as much glucose as willows of the same genotype grown in sheltered sites. Scientists are currently exploring the possibility of growing and harvesting these trees specifically for that purpose.

How the poplar packed away

It's a sobering thought that 75 per cent of the Earth's land surface has been impacted by human activity, often adversely, and there are few signs that that percentage will do anything other than increase in the foreseeable future.

Large amounts of the land in question have been contaminated by substances that are potential or actual hazards to health or the environment, either on the surface or below ground. In many cases, even when the harmful activity has ceased, its ill effects remain. Among the most damaging residues are from mining, oil, chemical and manufacturing industries, waste disposal and biological pollution sites. Discarded electronic equipment, pharmaceuticals, refrigerants, antibiotics, heavy metals such as lead and arsenic, and radioactive waste can all be land pollutants. Often mixtures of waste produce hazardous new cocktails of contaminants. Fortunately, contaminated land can be excavated and chemically treated to clean it up and returned to the site. While this produces quick results, which may suit urgent building projects, it is an expensive solution.

*Poplars have become renowned for
their capacity to grow healthily on
contaminated land.*

Some plants, fungi and microorganisms can thrive in settings that would damage or destroy other species. The initial findings were with plants that could absorb nickel. The most spectacular of these was the very rare *Pycnandra acuminata* tree from New Caledonia in the southwest Pacific, which produces a bluey-green sap containing a staggering 25 per cent nickel citrate. Unfortunately, its presence acted as an indicator in the past of where to dig for nickel.

Since the 1970s several wild and cultivated plants have been identified that can compartmentalize a range of heavy metals, such as lead and cadmium, and thus avoid being harmed by them. The process is known as phytoextraction, and the species that can perform it are known as hyperaccumulators.

In the early 1990s researchers experimented with hyperaccumulator species grown on contaminated land as a biological means of cleaning it up, a process known as phytoremediation. One of the first species to be used for this purpose was a hybrid poplar. This tree grows quickly, has a deep-spreading root system, readily takes up certain contaminants from soil and ground water, and has large leaves to contain them. The biomass absorbed by the poplar was then harvested and taken away to be processed, thus reducing the level of contamination on the land.

The best place to find suitable hyperaccumulator species may be from plants already growing on contaminated land. Ferns remain on contaminated mining sites after other plants have died out because they are capable of extracting

The Chinese brake fern: 'brake' is a corruption of 'bracken'.

and compartmentalizing arsenic and other heavy metals.
The Chinese brake (*Pteris vittata*), a fern indigenous to Asia,
southern Europe, tropical Africa and Australia, successfully
cleared land contaminated with arsenic from explosives
manufacturing after several years of harvesting fronds from the
ferns growing there. On contaminated sites the plant is usually
colonized by mycorrhizal fungi – the combination of the two
appears to increase efficient uptake of contaminants.

The success of other hyperaccumulators is variable. After the
1986 Chernobyl nuclear disaster in Ukraine, sunflowers
(*Helianthus*) grown on rafts were effective at extracting
strontium and caesium from contaminated water, but less
proficient at extracting the radioactive metals from the soil.
The same flower species was used in the cleaning-up operation
after the 2011 Fukushima nuclear disaster in Japan, but the
variety chosen had no phytoextraction capabilities, so didn't work.

Some pine trees can live for more than 1,000 years.

How the pine tree pined

The history of the world is recorded in the rings that grow year by year in the trunks of our trees: these are wider in the good times (warm weather, sufficient rain) and thinner in the bad ones (years with climate extremes).

Some trees live fast and die young: many hybrid poplar and birch trees last only around 30 to 60 years. At the other end of the scale, some of the pine trees and yew tree species growing today were well established before the dawn of written history.

Pioneer trees – those that appear early in forest clearings – tend not to survive for long because a shortage of nearby foliage means that pests and diseases focus on them. Meanwhile, they themselves have fast metabolisms that go for growth rather than generating chemical defence compounds.

By contrast, ancient trees have slow metabolisms: their leaves may survive for up to five years on very slowly extending shoots. Ironically, another poplar species, the quaking aspen (*Populus*

Some hybrid poplars grow very quickly, reaching up to 18 m (60 ft) in six years, but live no more than 60 years.

tremuloides), found in some cooler areas of North America, readily produces new stems from very ancient root stocks, thus surrounding itself with stands of clonal trees, some of which, though young, share a common ancestor of remarkable age: known as Pando (Latin for 'I spread'), a stand of male quaking aspens in Fishlake National Forest, Utah, USA, is estimated to be around 80,000 years old.

Some broad-leaf species, such as oak trees, live up to 1,000 years. Evergreens, such as the olive tree (*Olea europaea*), and conifers, such as the yew (*Taxus baccata*), may be twice as old, although it is difficult to tell for certain because the original trunks have often rotted away.

Bristlecone pines (*Pinus longaeva*) on windswept slopes on the higher mountains of California, Nevada and Utah exhibit the squat growth of ancient trees. They tend to lose their tops, which they can't support in old age, and produce new branches from their lower parts instead. This tendency can also be seen in other species, such as ancient oak trees. Bristlecone pine trees living a more protected life at lower elevations grow taller but die younger. At around 4,800 years of age, a US bristlecone pine named Methuselah was considered to be the world's oldest living tree. However, there is a continuing quest to find even older trees – in 2023 scientists found a Patagonian cypress or alerce (*Fitzroya cupressoides*) in southern Chile named Lañilawal that may be more than 5,000 years old. Alerce are already known to live up to 3,800 years of age. Some traditional houses in the Chilean Lake District are made from it, but it is unsustainable because it grows so slowly. The reason for the resilience of the wood in such buildings is that long-lived trees produce defence compounds – resins and waxes, tannins in oak trees and terpenoids in conifers – rather than putting on rapid growth. Such compounds accumulate with age and make it more difficult to propagate these trees vegetatively. Despite this, new trees from the oldest yews in Europe and from the Major Oak (a 1,000-year-old growth in Sherwood Forest, UK) have been propagated in Britain. In the USA, the Archangel Ancient Tree Archive uses cloning technology to replant ancient trees.

Growing conditions for ancient trees in woodlands and in the open are quite different. Trees in the open can grow a full canopy of branches, whereas those in a forest lose their lower branches in the competition for light. However, ancient trees that lose their top branches cannot compete with other forest trees and die out due to lack of light. By contrast, those located in the light in parkland can carry on growing.

The point the pineapple tried to make

Photosynthesis – the conversion of light into chemical energy – takes place in most plants, but the process is not the same in all species, as many have had to adapt to local conditions.

In order to photosynthesize, plants generally absorb carbon dioxide through pores in their leaves during daylight hours. They then combine the gas with water to create oxygen, some of which is released back into the air, and energy in the form of glucose.

That is broadly how the process works in normal conditions, but conditions are not always normal. Many plants have to compensate for varying quantities of water and sunlight. In particularly arid conditions, they may photosynthesize during the day but exchange gases only at night. They do this by keeping the stomata in their leaves shut during the day to minimize water loss in the heat, and opening them at night when it's cooler to absorb carbon dioxide.

Pineapples have developed a method of photosynthesis that is adaptable to a wide range of climatic conditions.

This variation on the basic photosynthesis theme was first observed in the Crassulaceae family of succulents, and is consequently known as crassulacean acid metabolism (CAM). CAMs are not of a type: they have evolved separately and differently in up to 35 different plant family groups.

The best known and most widely distributed plant that uses a CAM system is the domesticated pineapple (*Ananas comosus*). It is grown in more than 80 tropical and subtropical countries on around a million hectares (2.5 million acres) of land, and is the third most important tropical fruit crop in the world after bananas and mangoes.

Pineapples of the type on sale in shops today were domesticated around 6,000 years ago from their wild antecedent *Ananas comosus* var. *ananassoides*, which is drought-resistant and grows in savannahs and low shady forests on soils with low water-holding capacity.

The ways in which pineapples resemble and differ from their relatives the Poaceae – a family of grasses, such as rice and sorghum, from which they diverged around 100 million years ago – helps scientists to compare and contrast their respective photosynthesis systems. From this it has been learned that pineapples can restrict their water loss by using more than one CAM system: they can adapt to variations in light and humidity.

Other drought-tolerant crop plants with the CAM mechanism include opuntia, cacti, agave and aloe. *Opuntia* (prickly pear) is by far the most important agricultural cactus crop. Its pads are

The prickly pear is native to the Americas, but has become established as an invasive species in other parts of the world.

harvested and eaten as a sliced green vegetable and as a source of animal fodder. Its fruits are also valued, and cochineal dye is produced from the beetles that feed on it.

Cacti have skeletal structures that have been used as building material. There is great diversity of fruiting types in terms of size, shape, sweetness, sharpness and seed content. Agaves are grown for their fibre and used in sweeteners, beverages and food. They are also popular as ornamentals. Aloes are used for drinks, health foods and in cosmetics; some species can grow on saline soils and even be effectively irrigated with salt water. These under-researched species may provide some future solutions as resilient crops in response to climate change.

Despite the attractive appearance of buddleia, in many countries it is considered an invasive species.

How the buddleia grew up the wall

All over the world, a summer's day can be brightened up and the spirits raised by the sight of butterflies flitting around and settling to drink nectar from the beautiful and aromatic tapered flower heads of the *Buddleja* species.

The genus *Buddleja* is named in honour of Adam Buddle, an 18th-century English vicar and amateur botanist. The penultimate letter of its Latin name was an error by the Swedish taxonomist Carl Linnaeus, who mistook an 'i' for a 'j', but his spelling has been retained ever since in scientific writing, although the common English spelling is 'buddleia'.

Buddleja davidii – the most popular and widely cultivated of around 90 species – is also known as summer lilac, butterfly bush or orange eye. The second part of the binomial is a tribute to the 19th-century French missionary and naturalist Père Armand David. The plant is native to China and Japan but can grow just about anywhere and is now well established globally. It is a pioneer species that flourishes in fallow land and open river valleys, from where it will spread uphill until its progress is halted by shadier wooded areas. It can also take hold in the walls of buildings. Although the ease with which it proliferates around the world can make it an invasive alien species, the threat to native flora and the potential damage it can cause to the fabric of buildings are sometimes overlooked because of the beauty of the species' blooms. In Britain, for example, it escaped from gardens in the 1930s and settled first in limestone quarries before proliferating widely. In the 1950s it brightened up the generally dismal postwar landscape and became known as 'the bombsite plant', even though the term did scant justice to its capacity to thrive almost anywhere: it can send deep roots down to grow in the mortar of derelict buildings and in tiny crevices in their walls, as well as alongside roads, rivers and railway lines.

The hardiness and adaptability of buddleia have negative consequences that may sometimes outweigh its attractive appearance. In Britain it is today the second most invasive species after Japanese knotweed (*Reynoutria japonica*), which was also originally introduced as an interesting garden exotic. (Gardeners at first covered Japanese knotweed with iron frames because they thought it was delicate and vulnerable; how wrong they were!) Both species are good illustrations of the axiom that there is no such thing as a weed, there is only a plant in the wrong place.

Buddleja davidii is rampant not only in Britain: it is now found everywhere with a Mediterranean, temperate, subtropical or continental climate. Throughout mainland Europe, North and South America, Australia and New Zealand it has spread in the same way: from gardens to the wild. Its aversion to extremes of temperature has so far kept it out of the coldest parts of Alaska and northern China and the world's hottest steppes and deserts, but that may change with global warming.

So why is the buddleia so successful outside its natural habitat? Most animals leave its bushes alone; it has few pests that lay eggs on it or use it as a food source for caterpillars; few animals browse its leaves; only humans uproot it or spray it. It grows vigorously and reaches flowering age early in its life. Although its seeds survive in the ground for only just over two years, it has a very efficient seed dispersal mechanism: after flowering, the seed capsules mature over winter, drying out in the spring and releasing up to 3,000,000 seeds per plant over 10 m (32 ft) by wind dispersal, often aided by the vortex created by passing road and rail vehicles and water currents. Its plants also readily reproduce asexually from stem and root fragments. Buddleia bushes provide a great habitat for pollinating insects on waste ground and their nectar is highly attractive to butterflies. Buddleia is outstandingly proficient at perpetuating itself; so good, in fact, that there may be no stopping it. Efforts have been made in the USA to ban the sale of wild forms of buddleia and to breed completely sterile garden-worthy forms, but with variable success.

The rhododendron and the azalea tale

Humans have developed methods of grouping or separating plants since time immemorial to identify food, medicine, poisonous and ornamental plants.

Since Linnaeus introduced binomial scientific naming (see page 182), different scientific approaches to naming have evolved. Some scientists are 'lumpers' who group species together overlooking small differences that separate groups – they regard differences as less important than similarities. Their outlook is based on plant species evolving to adapt to varying environmental conditions. By contrast, 'splitters' are scientists who separate similar plants, highlighting significant differences to categorize them as separate groups. In some cases, splitting species up has been practised as the only way anonymous botanists could get their work recognized in print.

In the case of the azalea and the rhododendron, azaleas are now classed as a subfamily of rhododendrons; they are both in the genus *Rhododendron*. But most rhododendrons are evergreen;

This rhododendron was once classified as different from an azalea, but now the two plants are parts of a single genus.

azaleas are mainly deciduous except for Japanese azaleas (*Rhododendron japonicum*). Azaleas are generally smaller in stature than rhododendrons and have smaller flowers and leaves. While most rhododendrons are large – up to tree size, with big flowers and leaves – alpine species are small. Rhododendrons generally have ten or more stamens, while azaleas have only five. Azaleas have tubular or funnel-shaped flowers, whereas the flowers of rhododendrons are bell-shaped. Significantly, rhododendron flowers are held in round clusters of flowers while azaleas are generally single. Rhododendrons prefer dappled shade; most azaleas are more tolerant to lighter aspect and soil conditions, although again that is not the case with Japanese azaleas.

The names of plant species are changed when variation in a species is too great to be maintained as one species, or if an older name for a species is discovered. Modern genetics and statistical analysis are also used to determine if physically diverse specimens are a single species. Any proposed name change needs to be checked by competent taxonomists to ensure that it is valid. When names are changed, it is a hassle for everyone, including taxonomists, to remember obscure plant names and so there is still a tendency on occasions to recall the old name. In the case of well-known species, they tend to keep their old name – sometimes the scientific name becomes a common name. A case in point surrounds the species that Linnaeus named *Amaryllis belladonna*. Over the following 200 years, doubts emerged about whether the specimen he had first described was the species originating in South Africa or one of the similar species originating in Latin America. After much study and heated debate, it was eventually ruled in 1987 by the 14th International Botanical Congress that the South African species was the one originally described and that it is the true amaryllis. The New World species and cultivars sold all over the world to provide winter cheer from their showy flowers should henceforward be classified as *Hippeastrum*. Nevertheless, they are still commonly marketed as amaryllis as the name has stuck with the public. The likelihood of future large name changes is reduced in well-studied areas, but there are still many species to discover in some areas of the globe, including the tropics.

There is a world shortage of trained taxonomists due to underfunding in recent years, leading to greater reliance on citizen science to record new plant species. Taxonomy has undergone major changes to increase online public and scientific accessibility, allowing global contributions and collaboration based on the world's herbarium collections and digital photographs of plants with GPS data which allow unexplored areas to be identified. Science writers have faith that if small children can pronounce the names of the world's leading football stars and dinosaurs, they can learn plant names as long as the stories about interesting plants from around the world are exciting enough!

It is not hard to see how the bottlebrush plant got its name.

How the bottlebrush cleaned the bottle

Perhaps disappointingly, the bottlebrush is no use whatsoever for cleaning out bottles. It's just one of several plants that are named for their fortuitous resemblance to commonplace objects.

Bottlebrush is the common name for *Callistemon*, a genus of shrubs in the family Myrtaceae. A native of Australia, it is a bush that flowers in a wide range of striking colours.

Although this species has no known practical uses, many other plant varieties have been adopted as utensils by diverse creatures: chimpanzees carry sticks to test the depth of water and to catch termites; Galapagos woodpecker finches forage for food with cactus spines in their beaks.

But it's humans who throughout history have used plants most extensively to assist them in manual tasks. The practice of weaving plant fibres into cords and baskets is known to have been established 26,000 years ago and is almost certainly much

The outer casing of the gourd is so tough that it can be hollowed out and used as a liquid container.

older than that. Wooden spoons were first developed towards the end of the Palaeolithic Period (up to around 10,000 years ago). The earliest wooden knife handles are about 7,000 years old. In China, twigs evolved into chopsticks made from wood or bamboo around 5,000 years ago, and their use gradually spread throughout Southeast Asia.

One of the earliest domesticated plants is the calabash or bottle gourd (*Lagenaria siceraria*). Part of the *Cucurbita* genus, calabashes have been used as liquid containers since time immemorial. Although native to Africa, they were established in North and South America long before the 15th-century voyages of Christopher Columbus: its seeds have been identified in the fossilized dung of mastodons, an American species that became extinct around 11,000 years ago. There has consequently been much speculation about how calabashes made it to the New World in the first place: they are thought unlikely to have been imported by early travellers over what was once a land bridge between Russia and Alaska because the seeds wouldn't have survived the cold conditions of the last Ice Age; in any case, there's no archaeological evidence to support this explanation. However, it has recently been demonstrated that bottle gourds are capable of germinating after extensive periods floating on water, so it looks most probable that they crossed the Atlantic Ocean that way.

Luffa gourds (*Luffa aegyptiaca* and *Luffa acutangula*) – tropical and subtropical vines in the same family as the calabash, Cucurbitaceae – have two distinct human uses. Their young fruits are edible and popular as vegetables in India, China, Bangladesh and Vietnam. Mature fruits can be heat-dried, after which they become a straw-coloured abrasive (loofah) that is used for scrubbing saucepans in the kitchen and backs in the shower.

Internet resources

1. Gorse

beespoke.info/2014/01/20/blooming-gorse

workingforwildlife.co.uk/nectar-robbing-bees

2. Campanula

botsocscot.wordpress.com/2020/05/16/polyploid-harebells-a-commentary-on-julia-wilsons-recent-paper

plantsrescue.com/posts/campanula-isophylla

3. Ranunculus

youtube.com/watch?v=LOqrFBbx41I&t=1605s

4. Leopard lily

news.bbc.co.uk/earth/hi/earth_news/newsid_8108000/8108940.stm

botanic.cam.ac.uk/deceptive-daisys-ability-to-create-fake-flies-explained

5. Bergenia

guinnessworldrecords.com/world-records/627895-largest-banana-species

atlasobscura.com/articles/leaf-size-air-temperature-model-overheating-freezing

6. Miscanthus

awatrees.com/2013/01/06/psithurism

bbc.com/travel/article/20190424-the-english-vegetable-picked-by-candlelight

7. Amaryllis

pza.sanbi.org/amaryllis-belladonna

oldhousegardens.com/GloriaMundiHyacinth

8. Lettuce

labroots.com/trending/genetics-and-genomics/20268/history-lettuce-domestication-told-dna

9. Alfalfa

hal.science/hal-01216251/document

sciencedirect.com/topics/agricultural-and-biological-sciences/alfalfa

10. Crab-apple

primrose.co.uk/blog/plants/apple-tree-pollination

11. Catmint

rhs.org.uk/plants/nepeta

12. Buttercup

aber.ac.uk/en/news/archive/2009/june/title-77794-en.html

13. Rose

nature.com/articles/s41438-021-00689-7

14. Snowdrop

blog.edvotek.com/2021/03/02/the-science-of-snowdrops

15. Tulip

rhs.org.uk/plants/tulip

16. Bluebell

indefenseofplants.com/blog/tag/arbuscular+mycorrhizae

17. Hosta

scientificamerican.com/article/how-monarch-butterflies-evolved-to-eat-a-poisonous-plant

18. Hydrangea

rootingforblooms.com/why-are-my-hydrangea-leaves-turning-brown-black-white-red-and-more

19. Camellia

rhs.org.uk/plants/camellia/frequently-asked-questions

20. Passionflower

kew.org/read-and-watch/plant-seed-dispersal-animal-poo

21. Daffodil

a-z-animals.com/blog/narcissus-vs-daffodil-is-there-a-difference

22. Honeysuckle

hannahlongmuir.co.uk/the-magic-of-dusk

23. Curry plant

kew.org/read-and-watch/why-do-plants-smell

phys.org/news/2015-07-heaven-scent-fragrance-roses.html

24. Pansy

ncbi.nlm.nih.gov/pmc/articles/PMC7817606

nature.com/scitable/knowledge/library/plant-resistance-against-herbivory-96675700

25. Musk

rhs.org.uk/plants/10797/malva-moschata/details

26. Fern

nhm.ac.uk/our-science/our-work/biodiversity/evolution-fern-diversity.html

27. Acacia

gardeningknowhow.com/ornamental/trees/acacia/acacia-tree-types.htm

28. Peony

rhs.org.uk/plants/peony

29. Chrysanthemum

chrysanthemumsvancouver.com/photoperiodic.html

30. Sweet pea

biodiversitylibrary.org/page/34071946#page/16/mode/1up

dabahdesigns.com/nj-landscape-design-blog/tag/Thigmomorphogenesis

31. Dahlia

loc.gov/everyday-mysteries/botany/item/what-causes-flowers-to-have-different-colors

thegardenstrust.blog/2022/03/12/violets-are-red-and-roses-are-blue

32. Dandelion

seas.harvard.edu/news/2012/08/uncoiling-cucumbers-enigma

bioexplorer.net/thigmotropism.html

33. Orchid

ncbi.nlm.nih.gov/pmc/articles/PMC4549959

34. Arum

loc.gov/everyday-mysteries/botany/item/what-is-the-largest-flower-in-the-world

35. Aubretia

thespruce.com/aubrieta-profile-4688622

36. Poppy

plantura.garden/uk/flowers-perennials/red-poppy-overview

37. Apple
rhs.org.uk/problems/fruit-biennial-bearing

38. Petunia
nhm.ac.uk/discover/news/2021/september/plants-and-pollinators-use-electric-fields-to-find-each-other.html

39. Sunflower
vcresearch.berkeley.edu/news/how-sunflowers-follow-sun

40. Daisy
blogs.memphis.edu/biology/2019/07/13/daisies-and-dinosaurs

41. Oak
thesmarthappyproject.com/galls

42. Houseleek
getplanta.com/article/trivia/commonhouseleek

43. Celandine
ecosystemgardening.com/most-hated-plants-lesser-celandine.html#google_vignette

44. Willow
natgeos.com/types-of-willow-trees

45. Poplar
frontiersin.org/articles/10.3389/fpls.2020.559059/full

ancienttreearchive.org

46. Pine tree
forestlegality.org/risk-tool/species/alerce

sdnhm.org/oceanoasis/fieldguide/popu-tre.html

47. Pineapple
sciencedaily.com/releases/2015/11/151102125733.htm

48. Buddleia
plantright.org/watch/buddleja-davidii

49. Rhododendron/Azalea
almanac.com/plant/rhododendrons-and-azaleas

50. Bottlebrush
science.org/content/article/scientists-solve-mystery-world-traveling-plant

Index

About the author

DR ANDREW ORMEROD is a botanist who holds an honours degree in agricultural botany and a doctorate in plant breeding. He has held positions at Reading University and Unilever Research and as an Honorary Research Fellow at the University of Kent. He worked at the Eden Project for 15 years, from when it first opened in 1998, and where he was involved in plant procurement for the exhibits. He lectures, gives tours and is the author of two plant blogs, cornucopiaalchemy and Cornwall and South West Fruit Focus.

Acknowledgements

I would like to thank those who I contacted during research for the book to check facts including Christian Körner, Beverley Glover, Robin Clery, Kelsey Byers, Chris Warner, Alistair Culham and Julia Wilson for their suggestions.

First published in Great Britain in 2024 by

Greenfinch
An imprint of Quercus Editions Ltd
Carmelite House
50 Victoria Embankment
London EC4Y 0DZ

An Hachette UK company

A CIP catalogue record for this book is available from
the British Library.

Hardback ISBN 978-1-52943-055-4
Ebook ISBN 978-1-52943-056-1

10 9 8 7 6 5 4 3 2 1

Text by Dr Andrew Ormerod
Design by John Round Design
Cover and interior artwork by Romy Palstra

Printed and bound in China

Papers used by Greenfinch are from well-managed forests and other
responsible sources.